# RADIATION PROTECTION

## Recommendations of the International Commission on Radiological Protection

ICRP PUBLICATION 4

# Report of Committee IV
(1953–1959)

*on*

# Protection Against Electromagnetic Radiation Above 3MeV and Electrons, Neutrons and Protons

(*Adopted* 1962, with revisions adopted in 1963)

PUBLISHED FOR

The International Commission on Radiological Protection

BY

PERGAMON PRESS

OXFORD . LONDON . EDINBURGH . NEW YORK
PARIS . FRANKFURT

1964

T0333910

PERGAMON PRESS LTD.
Headington Hill Hall, Oxford
4 & 5 Fitzroy Square, London, W.1

PERGAMON PRESS (SCOTLAND) LTD.
2 & 3 Teviot Place, Edinburgh, 1

PERGAMON PRESS INC.
122 East 55th Street, New York 22, N.Y.

GAUTHIER-VILLARS ED.
55 Quai des Grands-Augustins, Paris, 6e

PERGAMON PRESS G.m.b.H.
Kaiserstrasse 75, Frankfurt am Main

Distributed in the Western Hemisphere by
THE MACMILLAN COMPANY · NEW YORK
pursuant to a special arrangement with
Pergamon Press Limited

Library of Congress Catalog Card Number 63–22751

Set in 'Monotype' Baskerville 10 on 11 pt.
and printed in Great Britain by
Bell & Bain, Ltd., Glasgow

# CONTENTS

## REPORT OF COMMITTEE IV

PREFACE

REPORT

A. Introduction

B. RBE and $QF$

C. Permissible Flux Densities
   (a) Neutrons and Protons Excluding Thermal and Intermediate Energy Neutrons
   (b) Thermal and Intermediate Energy Neutrons
   (c) Electromagnetic Radiation Above 3 MeV and Electrons

D. Principles Regarding Working Conditions and Monitoring
   (a) General Principles
   (b) Monitoring

E. Recommendations on Equipment and Operating Conditions
   (a) General Considerations
   (b) Installations (Structural Details)
   (c) Installations (Operational Details)

APPENDICES
   I. Ion Density and Linear Energy Transfer
   II. Absorption Data for Electromagnetic Radiation
   III. Particle Ranges
   IV. Absorption Data for Neutrons

LIST OF REFERENCES

### RECOMMENDATIONS OF THE COMMISSION

Organization
A. Prefatory Review
B. Basic Concepts
C. Maximum Permissible Doses
D. General Principles Regarding Working Conditions

# PREFACE

THIS report was commenced under the Chairmanship of Prof. W. V. Mayneord, and a draft was completed by the 1956–1959 Committee under the Chairmanship of Prof. H. E. Johns.

The co-operation of Dr. Bo Lindell and Dr. H. H. Rossi, as members of the 1958 Publication Committee, in the revision of the original draft, is gratefully acknowledged. During final editing, undertaken by Dr. B. M. Wheatley, the report was reviewed by the 1959–62 members of the Committee (see page 24).

This publication also contains a reprint of the Recommendations of the International Commission on Radiological Protection.

MEMBERSHIP OF COMMITTEE IV
*(Protection against electromagnetic radiation above 3 MeV and electrons, neutrons and protons)*
1953–1956

| | |
|---|---|
| W. V. MAYNEORD, *Chairman* | (Great Britain) |
| L. H. GRAY | (Great Britain) |
| H. E. JOHNS | (Canada) |
| H. W. KOCH | (U.S.A.) |
| P. LAMARQUE | (France) |
| J. S. LAUGHLIN | (U.S.A.) |
| J. S. MITCHELL | (Great Britain) |
| B. MOYER | (U.S.A.) |
| C. A. TOBIAS | (U.S.A.) |
| F. WACHSMANN | (Germany) |

1956–1959

| | |
|---|---|
| H. E. JOHNS, *Chairman* | (Canada) |
| J. S. MITCHELL, *Vice-Chairman* | (Great Britain) |
| L. H. GRAY | (Great Britain) |
| F. HERČÍK | (Czechoslovakia) |
| G. JOYET | (Switzerland) |
| W. H. KOCH | (U.S.A.) |
| J. S. LAUGHLIN | (U.S.A.) |
| W. V. MAYNEORD | (Great Britain) |
| C. A. TOBIAS | (U.S.A.) |
| M. TUBIANA | (France) |
| F. WACHSMANN | (Germany) |

# REPORT OF COMMITTEE IV

ON

# PROTECTION AGAINST ELECTROMAGNETIC RADIATION ABOVE 3 MeV AND ELECTRONS, NEUTRONS AND PROTONS

## A. INTRODUCTION

(1) The maximum permissible doses which are recommended for external radiation are given in Section 4 of ICRP Publication 6, which is reprinted on pages 26–44 of this report.

(2) Although the hazards associated with the use of high energy X-rays and heavy particles (including neutrons and protons) are in many ways analogous to those occurring with X-rays produced at lower voltages, there are special risks and circumstances which call for detailed separate discussion. These problems are, therefore, considered below. The scope of this report covers electrons, electromagnetic radiation with quantum energies above 3 MeV, and neutrons and protons with energies up to 1000 MeV.

(3) It will be noted that the lower energy limit of electromagnetic radiation considered in this report is 3 MeV. This value was regarded by the International Commission on Radiological Units and Measurements (ICRU) as the upper limit of the energy range over which the röntgen should be used. It therefore provided a logical division between the scope of Committees III and IV. The point of separation is also convenient in eliminating from our consideration the problems of gamma-rays from most radioactive substances. Nevertheless many of the general recommendations in the report of Committee III* are also applicable to the higher energy radiations, and the reader is therefore referred to this report.

## B. RBE AND QF

(4) All doses specified in the Commission's recommendations are expressed in rems,† which implies that the absorbed dose (expressed in rads) should be multiplied by an appropriate weighting factor. Previously the weighting factor was termed the " RBE ". The use of this term both in radiobiology and for protection purposes presents certain problems, which are discussed in detail in the Report of the ICRP/ICRU Committee on RBE (*Health Physics* **9**, No. 4, (1963)).

(5) The Commission now endorses the following statement, taken from Report 10a[1] of the International Commission on Radiological Units and Measurements (ICRU) :

> The term " RBE dose " has in past publications of the Commission not been included in the list of definitions but was merely presented as a " recognized symbol ". In its 1959 report the Commission also expressed misgivings over the utilization of the same term, " RBE ", in both radiobiology and radiation protection. It now recommends that the term RBE be used in radiobiology only and that another name be used for the linear-energy-transfer-dependent factor by which absorbed doses are to be multiplied to obtain for purposes of radiation protection a quantity that expresses on a common scale for all ionizing radiations the irradiation incurred by exposed persons. The name recommended

---

* ICRP Publication 3. Report of Committee III on Protection Against X-Rays up to Energies of 3 MeV and Beta- and Gamma-Rays from Sealed Sources (1960).

† The term " dose ", when used in this context, is now to be understood as " dose equivalent " (see discussion below).

for this factor is the *quality factor* (*QF*). Provisions for other factors are also made. Thus a distribution factor (*DF*), may be used to express the modification of biological effect due to non-uniform distribution of internally deposited isotopes. The product of absorbed dose and modifying factors is termed the *dose equivalent* (*DE*). As a result of discussion between ICRU and ICRP the following formulation has been agreed upon :

*The Dose Equivalent*

1. For protection purposes it is useful to define a quantity which will be termed the " dose equivalent " (*DE*).
2. (*DE*) is defined as the product of absorbed dose, *D*, quality factor (*QF*), dose distribution factor (*DF*) and other necessary modifying factors.
$$(DE) = D(QF)(DF) \ldots$$
3. The unit of dose equivalent is the " rem ". The dose equivalent is numerically equal to the dose in rads multiplied by the appropriate modifying factors.

(6) With regard to the actual values of *QF* that should be used for radiation protection calculations, the Commission endorses the " RBE " values which it published in 1955.[2] These values are related to the LET of the radiation independently of other exposure factors. It is recommended that with regard to specification of radiation quality the basic parameter be LET$_\infty$ (the " stopping power "), defined as the energy loss per unit distance of the charged particles originally set in motion by electromagnetic radiation or neutrons, or of the charged particles which originate in radiation sources (alpha-rays, beta-rays, etc.), i.e. the delta-rays are not counted as separate tracks. The *DE* (expressed in rems) is obtained by summation of the products of doses delivered at any LET and the appropriate *QF*-factors, as well as any other factors recommended by the Commission, for example the " *n* " factor. Simplifications of this procedure are allowed provided they do not result in an underestimate of the true *DE*. An example of such a simplification is the use of a single value of *QF* for all fast neutrons.

(7) The relationship between *QF* and LET recommended for radiation protection calculations is the following :

TABLE I. LET–*QF* RELATIONSHIP

| LET$_\infty$ (keV per micron in water) | *QF* |
|:---:|:---:|
| 3.5 or less | 1 |
| 3.5–7.0 | 1–2 |
| 7.0–23 | 2–5 |
| 23–53 | 5–10 |
| 53–175 | 10–20 |

(8) In practice the *QF* for X- and gamma-rays is taken as unity, and for electrons it is only greater than unity at very low energies.

(9) Most practical *DE* problems consist in the evaluation of the hazard due to a mixture of neutrons and gamma radiation. The *QF* of neutrons as a function of neutron energy has been evaluated for neutron energies up to 10 MeV.[3] If the neutron energy distribution is known, the absorbed dose due to neutrons may then be multiplied by an appropriate *QF* to obtain the *DE*. If the precise neutron energy is unknown, the absorbed doses due to neutrons and gamma-rays should be evaluated separately. The sum of the neutron doses multiplied by 10 and the gamma-

ray doses multiplied by 1 may be considered an upper limit of the *DE*. Finally, the simplest approach is merely to measure the total absorbed dose and to multiply it by a *QF* of 10. While being the simplest, this method may result in an overestimate by a factor that can approach 10.

(10) For heavy recoil nuclei the LET may be greater than 175 keV/$\mu$. There is, however, experimental evidence that even in this case the *QF* probably does not exceed 20. This value is therefore considered to be appropriate for heavy recoil nuclei.

(11) At high energies, values obtained in experimental radiobiological investigations[4,5] may not be directly applicable to protection problems because of the additional consideration of physical differences of the dose and LET distribution in whole body exposure and the exposure of small specimens, and also because of differences in the order of magnitude of the dose levels in the two cases.

## C. PERMISSIBLE FLUX DENSITIES

### (a) NEUTRONS AND PROTONS EXCLUDING THERMAL AND INTERMEDIATE ENERGY NEUTRONS

(12) Over a wide energy range the absorbed doses of neutrons and protons are maximal at or near the body surface.[3] Some critical organs such as the lens of the eye, and the male gonads, are also near the surface, so that the recommended maximum permissible fluxes derived below are related to absorbed dose at the surface of the body. Under some exposure conditions, where certain limited parts of the body are exposed to a collimated beam of radiation, this criterion may be invalid. Exposure to maximum permissible flux may then result in less than the maximum permissible absorbed dose in the particular part of the body which is the critical organ for that individual exposure. Committee IV is aware that a single table of maximum permissible fluxes interprets the Recommendations of the Commission in a more restrictive way than tables embodying parameters for different types of limited or non-uniform exposure. However, present data on biological effects and neutron dosimetry do not allow more refined calculations to be carried out in a sufficiently reliable manner.

(13) On absorption in the body, primary particles will produce a number of secondary particles of lower energy. The local biological effect is therefore due to the sum of the effects of a number of particles reaching the point of observation from all directions. The LET of these secondary particles will, in general, be different from that of the primary particle and, of course, the LET of each particle will vary along its track. The *QF* therefore depends on the weighted mean of the LETs associated with the various particles contributing to the absorbed dose. Figure 1 gives the best available estimates of *QF* for neutrons and protons.[3,6]

(14) The correspondence between tissue dose and incident particle flux may be calculated by integrating the dose equivalents arising from the energy losses of all the particles within the volume of interest, usually taken to be near the surface of a phantom irradiated at normal incidence. Figure 2 shows particle fluxes corresponding to a dose rate of 2.5 mrem/h. The flux values are for primary particles in beams assumed to be in equilibrium with secondary particles as a result of passage through shielding. Exposure to these fluxes for 40 hours a week results in a dose of 5 rems in 50 weeks. Figure 2 presents the results of Snyder[3] at energies up to 10 MeV; at energies above 10 MeV the results of Neary and Mulvey[6] have been used. At high energies the mean free paths of the secondary particles are comparable with those of the primaries, and one may derive different values of primary flux corresponding to a given absorbed dose depending on whether or not secondary particles are already present in the incident beam. The most restrictive case will be that in which secondary particles are in equilibrium with the primary beam, for example, as a result of passage through shielding. Calculation shows that, provided the shielding is more than one mean free path thick, then the material of which it is constructed is of minor importance.

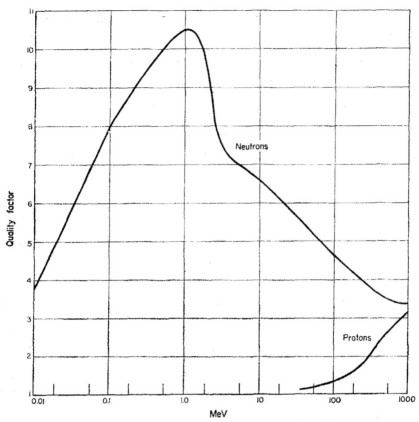

FIG. 1. Values of $QF$ for neutrons and protons of various energies. The curves are smoothed from tabulated data by Snyder[3] and Neary and Mulvey[6].

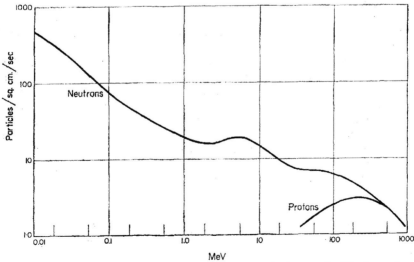

FIG. 2. Particle flux densities corresponding to a dose rate of 2.5 mrem/h. The flux values are for primary particles in beams assumed to be in equilibrium with secondary particles as a result of passage through shielding. The curves are smoothed from tabulated data by Snyder[3] and Neary and Mulvey[6].

#### (b) Thermal and Intermediate Energy Neutrons

(15) Neutrons of energy lower than those included in Fig. 2, impart most of their dose by neutron absorption. The two most important processes are the H $(n, \gamma)$ D and the $N^{14}$ $(n, p)$ $C^{14}$ reactions.

For convenience, Table II includes data for 0.02 MeV neutrons, whose energy is above the intermediate range, in order to overlap the data of Fig. 2.

TABLE II. VALUES OF $QF$ AND THE FLUX CORRESPONDING TO 2.5 MREM/H (AFTER SYNDER[3])

| Neutron energy (MeV) | Thermal | 0.0001 | 0.005 | 0.02 |
|---|---|---|---|---|
| $QF$ | 3 | 2 | 2.5 | 5 |
| Neutrons/cm²/sec corresponding to 2.5 mrem/h | 670 | 500 | 570 | 280 |

#### (c) Electromagnetic Radiation above 3 MeV and Electrons

(16) In interpreting the Recommendations of the Commission the Reports of ICRU[1,7] will be found useful. Photodisintegration processes may add a small contribution to the absorbed dose and may produce sufficient neutrons to contribute to shielding problems (see Appendix IV).

### D. PRINCIPLES REGARDING WORKING CONDITIONS AND MONITORING

#### (a) General

(17) The Recommendations of the Commission (reprinted on pages 26–44 of this report), particularly Section D, should be consulted.

(18) For each installation a radiation protection officer should be appointed whose duty it is to bring the hazards to the notice of staff and visitors and to see that the appropriate safety measures are carried out.

(19) Routine operation of an installation shall be deferred until a complete survey has been made and the installation has been found to be satisfactory.

#### (b) Monitoring

(20) The radiation protection officer shall require that all personnel associated with the operation of an accelerator shall be continuously monitored individually with a film badge, pocket dosimeter, or other appropriate device such as a neutron track plate for neutrons.

(21) No personnel shall enter an area in which a radiological hazard could conceivably exist until the area has been monitored and the radiation protection officer has specified appropriate working conditions.

(22) No single instrument exists which will measure all radiations dealt with in this report. It is therefore essential that a competent person shall decide which radiations can conceivably be produced in a given installation, and that he shall specify measuring instruments accordingly. His decisions should take account of abnormal or fault conditions as well as normal operating conditions. It should be remembered that the nature as well as the spectrum of the radiation may be changed by shielding materials.

(23) Where ionization chambers are used, the thickness and composition of the wall, and the nature and pressure of the gas filling, must be appropriate to the type of radiation being measured.

(24) All monitoring instruments should be calibrated, and some simple method of checking constancy should be devised. The calibration should be either specific to the conditions encountered

in practice, or should be in the form of an energy response curve from which the calibration factor appropriate to a specific condition may be derived.

(25) The techniques of measuring the various radiations covered in this report are numerous and reference should be made to the literature[3,7-11] to ensure that the monitoring equipment chosen is adequate and sufficient for the purpose.

(26) When appropriate, area monitoring should include measurements of airborne and contamination hazards arising from induced radioactivity. In some cases sufficient radioactivity may be induced in the equipment to produce an external radiation hazard, which should be assessed.

### E. RECOMMENDATIONS ON EQUIPMENT AND OPERATING CONDITIONS
#### (a) GENERAL CONSIDERATIONS
##### X-ray Generators

(27) Space and distance considerations.

The X-ray beam angulation should be restricted wherever possible. High energy X-rays are produced predominantly within a fairly narrow cone and secondary products of high-energy radiations are projected mostly in the forward direction. The distance of personnel from the X-ray target of an accelerator should be made as great as possible in order to take advantage of the reduction of the primary beam intensity with distance and thus to reduce the required thickness of the protective barriers. The beam intensity in any direction is approximately inversely proportional to the square of distance to the target.

(28) Practical protective materials.

Concrete is recommended as the most economical choice for permanent installations. Where space is at a premium and greater attenuation than that afforded by ordinary concrete is necessary, the concrete may be loaded with uniformly distributed material of higher atomic number (e.g. scrap steel, iron ore, barium rock).

(29) Thickness of shield required.

Shield thicknesses can be computed according to the method outlined in Appendix II.

##### Electron Generators

(30) Materials used for the shielding of accelerators producing beams of electrons should be of low atomic number to minimize the production of X-rays. Neglect of this precaution may give rise to an X-ray shielding problem of considerable magnitude. Concrete is recommended, from the standpoint of economy, efficiency and structural adaptability.

(31) It must be remembered that the dose rates in electron beams may be extremely high and lethal effects could be produced in small fractions of a second. Also, owing to the ease with which electrons are scattered, no room should be entered in which an electron beam exists.

##### Generators of Neutrons and Heavy Particles

(32) Attention is drawn to the fact that heavy particle accelerators also produce X-rays, and a special check should be made on this radiation.

(33) Hydrogenous materials are usually the most practical neutron shields. Concrete, water and wax are good moderators of fast neutrons as they contain a high percentage of hydrogen; concrete is generally found to be the most convenient for constructional reasons and because it is also a suitable material for X-ray shielding.

(34) Under no circumstances should a beam of particles be directly aligned by eye due to the danger of cataract production.

## (b) Installations (Structural Details)

(35) Primary protective barriers shall be provided for any area exposed to the beam. Attention is drawn to the advantages of siting a high energy accelerator so that its primary beam is directed below ground. Data given in Appendices II and IV may be of value in determining thicknesses of barriers required in particular circumstances.

(36) It is very desirable to impress on building contractors the necessity that concrete barriers be homogeneous and of the specified density.

(37) The shielding requirements on the door to the radiation room can be reduced by putting this door at the end of a maze leading to the control room.

(38) The area containing the accelerator controls shall be separate from the radiation room.

(39) Holes in the barrier for pipes, conduits and louvres shall be provided with baffles, so that the radiation transmitted does not substantially exceed that transmitted through the surrounding barrier.

(40) Windows and doors shall offer substantially the same degree of attenuation as that needed at adjacent parts of the barrier in which they are located.

(41) Lead, if it is necessary in such structures as doors, shall be mounted in such a manner that it will not " creep " because of its own weight. It should be protected against mechanical damage.

(42) Welded or burned lead seams are permissible, provided that the lead equivalent of the seams is at least as great as that required of the lead sheet.

(43) At the joints between different protective barriers the overlap shall be at least as great as the thickness of the thicker barrier.

(44) It should be remembered that the target is not the only source of radiation. In both electron and positive ion accelerators the beam usually strikes other objects such as vacuum chamber walls and electrode supports. In positive ion accelerators, secondary electrons travelling along the accelerator in the reverse direction will produce radiation as they strike the vacuum chamber walls and electrodes.

(45) Attention should be drawn to the possibility of radiation leaks around the apparatus where the protection has been cut away to insert experimental devices. Care should be taken to provide adequate protection against the radiation emitted from the target at large angles to the electron beam.

(46) The radiation protection requirements depend to a large extent upon the occupancy of surrounding areas. Therefore, the installation should be located as far as practicable from other occupied regions. If the useful beam is frequently directed towards the floor, the installations should preferably be located on the lowest floor with the earth directly below the floor of the radiation room. The protection requirements will be reduced by the use of a one-storey building since the shielding on the ceiling of the radiation room can be minimized. Further economy may be attained by restricting the movement of the beam.

(47) If possible, all personnel should be stationed at a considerable distance from the target, and in directions at large angles to the useful beam. If this is done, less structural protection will be required, because the primary radiation in these directions is lower in intensity. Also, the scattered radiation is less penetrating than at small angles. Advantage should also be taken of the shielding afforded by the structure of the accelerator itself.

(48) Barriers should be sufficiently thick to provide adequate shielding under all conditions (e.g. beam positions, target materials) in which the equipment can operate.

(49) All wall, floor and ceiling areas not provided with protective barriers for the primary beam shall be shielded with secondary protective barriers of thickness sufficient to reduce the attenuated and scattered radiation in occupied spaces below maximum permissible levels.

### (c) INSTALLATIONS (OPERATIONAL DETAILS)

(50) In addition to the radiation hazards associated with the useful beam during normal operation, others may exist. Measurements should be made to evaluate induced radioactivity, and unusual stray radiation fields. Significant levels of induced radioactivity may persist after ceasing accelerator operation. Parts of the equipment, especially the vacuum chamber and target, must be handled with adequate precautions. Activation of the air, resulting in the production of nitrogen-13, oxygen-15, or argon-41, may make it necessary to install special ventilation equipment. In some circumstances, such as when using tritium targets, the vacuum equipment and the exhaust air may become contaminated independently of the radiation output of the apparatus.

(51) Stray radiation in unexpected directions may be present when an accelerator is operated under abnormal conditions. Stray beams may constitute a radiation hazard due to some maladjustment of the control circuits, even when the useful beam is at zero intensity.

(52) The Committee wishes to draw attention to the existence of non-radiological hazards, and suggests that information on electrical, mechanical and toxic hazards be sought from national codes of practice. A toxic hazard, which may be significant if high output apparatus is operated with insufficient ventilation, arises from the production of noxious gases, including ozone and oxides of nitrogen.

### Installations for Medical Use

(53) Means should be provided for the continuous observation of, and oral communication with, the patient during treatment.

(54) The observation window should be so located that it cannot be exposed to the useful beam.

(55) A housing shall be provided so that the leakage radiation at a distance of 1 metre from the source does not exceed either 1 röntgen in an hour or 0.1 per cent of the useful beam dose rate 1 metre from the source, whichever is the greater.

(56) Permanent diaphragms or cones used for collimating the useful beam shall afford the same degree of protection as the tube housing. Adjustable or removable beam defining diaphragms or cones shall be constructed so as to reduce the integral dose to the patient as much as practicable. In no case shall they transmit more than 5 per cent of the useful beam.

(57) A transmission monitoring chamber in the useful beam is recommended for observing the radiation intensity. A reliable check of this monitoring equipment should be made regularly.

### Industrial Installations

(58) In view of the wide range of circumstances likely to be met in industrial installations, particular responsibility must fall upon the radiation protection officer to see that workers are aware of the dangers. Inasmuch as many industrial exposures will take place in large workshops in which heavy roofing for the equipment cannot easily be provided the danger of scattered radiations should not be overlooked.

(59) It is most important that monitoring facilities be provided to ensure that the doses received by the workers are not greater than the maximum permissible.

(60) The range of movement of the beam of radiation should be restricted as much as possible and the useful beam should as far as practicable not be pointed at the operator's room or at adjacent places where persons are working.

(61) The useful beam should be so directed that the radiation will be scattered at least twice before it traverses openings in secondary protective barriers.

(62) There are great advantages in the use of solid concrete blocks so designed that no continuous cracks extend through the barrier. In the useful beam no continuous crack should extend through more than 25 per cent of the barrier.

(63) Consideration should be given to the protection of workers (e.g. crane operators) who may occasionally be located directly above the radiation rooms. As an example interlocks should be provided to limit the crane operation to locations where a radiation survey indicates adequate protection.

## Research Installations

(64) Unusual radiation hazards may occur in research installations. It is essential that those working in the vicinity of a research installation be subject to adequate personal monitoring and that surveys of radiation levels around the machine be made. The radiation levels must be considered when determining the permissible working periods in particular places.

(65) A record of the times of operation and operating conditions of the equipment, including the direction of the beam, should be kept.

# APPENDICES

THE following appendices give data which may be found useful when considering $QF$ values or when designing shielding.

## APPENDIX I

### ION DENSITY AND LINEAR ENERGY TRANSFER

In Table I of this report $QF$ values were given as a function of the average linear energy transfer (LET) for ionizing particles. Although one can ascribe an average LET value to a particle of a given type with a given energy, it is well known that in reality the biological material bombarded by such a particle is exposed to a whole spectrum of LET values. The spectral distribution of these LET values gives more information than a single average LET value.

The determination of a complete LET spectrum for an ionizing radiation is by no means simple. Even in a simple case, for instance along the track of a heavy particle such as a proton, the LET varies because of the delta-rays produced at intervals along the track. Since LET values range from 0.1 to over 500 keV/micron the LET spectrum is most conveniently represented on a logarithmic scale $\lambda$ where $\lambda = \log_{10}(\text{LET})$. There is considerable uncertainty in the number of low energy delta-rays produced, and also in the appropriate LET for low energy electrons (100 eV). For this reason Howard-Flanders, who has reviewed the subject,[12] has represented all the energy dissipated by electrons of 1000 eV or less as a rectangle of appropriate area and made no efforts to show the distribution of LET values for these low energy electrons. His data are presented in Fig. 3.

The fraction of X-ray or beta-ray energy dissipated by delta-ray tracks of length (along the track) less than 100, 200, 500 or 1000Å is represented by areas within the delta-ray rectangles to the right of the points marked 1, 2, 3 and 4 respectively. Those areas may be deleted from the LET spectrum when treating data on materials for which it is thought that the short range delta-rays are without effect.

The top section of Fig. 3 shows the distribution for $Co^{60}$ gamma-rays. It is seen that LET values extend from the minimum possible value up to about 10 keV/$\mu$ and the delta-rays then produce tracks with LET values from 10 to 30 keV/$\mu$. In the middle section of the figure the LET distribution for 25 MeV X-rays is given, and it will be seen that its general shape is very similar to that for $Co^{60}$. The 220 keV X-ray beam also shows a broad spectrum of LET values.

The top section of the figure also shows the LET distribution for alpha-particles, whose original energy was 5.3 MeV. These produce tracks with LET values from 40 to 130 keV/$\mu$ while the delta tracks produced by them distribute their energy at lower LET values from 10 to 30 keV/$\mu$.

The bottom section of the figure shows the distributions for energetic hydrogen, helium, carbon and neon nuclei. The delta-ray spectrum given applies to all four particles. Thus the 200 MeV singly charged neon nucleus dissipates about 60 per cent of its energy with LET values between 300 and 500 keV/$\mu$ and the rest as delta-rays with values between 1 to 30 keV/$\mu$. This graph should be of value to those contemplating biological experiments with high energy beams. LET distributions have been independently calculated by Danzker, Kessaria, and Laughlin[13] for a number of X-ray spectra, using total electron spectra obtained with the methods of Spencer, Attix and Fano[14,15]. Comparison with the above results shows close agreement in the cases common to both.

10

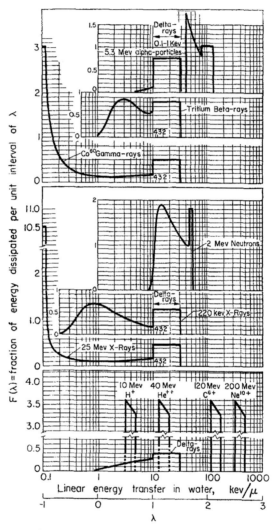

FIG. 3. The LET spectra in water for various commonly used ionizing radiations. $F(\lambda)$, the fraction of energy dissipated per unit interval of $\lambda$ is plotted as a function of $\lambda$ on a linear scale, where $\lambda = \log_{10}$ LET. For convenience, the horizontal scale is also marked directly in LET in keV/$\mu$ in water on a logarithmic scale. The fraction of energy dissipated between any two LET values is equal to the area under the curve between these limits. The rectangular areas represent the fraction of energy dissipated by delta-rays of energy 1000 eV and less. LET values for the core of the tracks and for delta-rays with energies above 100 eV are shown separately. These spectra apply as long as the track segments are shorter than 100Å. For track segments greater than 100Å, the right-hand side of the delta-ray spectrum should be cut short at the points marked 1, 2, 3, 4 as explained in the text. From Howard-Flanders[12].

APPENDIX II

ABSORPTION DATA FOR ELECTROMAGNETIC RADIATION

Narrow beams of monochromatic radiation are attenuated in an absorber according to an exponential law expressed by the following equation.

$$I/I_0 = \exp(-\mu d)$$

where $I$ is the intensity at the point of measurement

$I_0$ is the intensity at the same point but in the absence of the absorber

$\mu$ is the absorption coefficient of the absorber for the incident radiation

$d$ is the barrier thickness measured in units reciprocal to those used for expressing the absorption coefficient.

(i.e. if $\mu$ is the mass absorption coefficient in cm²/g then $d$ must be in g/cm², and if $\mu$ is the linear absorption coefficient in cm⁻¹ then $d$ must be in cm). One of the significant sources of error in using this equation is the uncertainty in the density of the material used, especially when using poured concrete.

Data for some common materials are given in Fig. 4 based on values published by Grodstein[16].

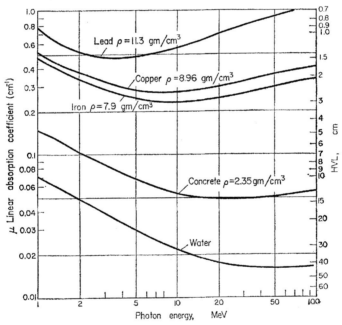

FIG. 4. Linear absorption coefficients and half value layers for common shielding materials. After Grodstein[16].

In practice few monochromatic beams are sufficiently narrow to satisfy the equation above, and radiation scattered by the absorber reaches the point of measurement. Thus the intensity or dose-rate of the beam reaching this point is greater than that expected from the equation by a factor known as the Buildup Factor. Buildup factors depend not only on the thickness and material of the absorber, but also on the geometry of the irradiation conditions. A uniformly radiating source completely surrounded by shielding would be represented by point isotropic geometry, whereas a wide beam normally incident on a wall would be represented by plane monodirectional geometry. Values have been tabulated by Goldstein and Wilkins[17], some of which are illustrated in Table III. It is clear that neglect of the buildup factors can introduce large errors.

The scattered radiation produced in the absorber has an extended spectral distribution, and, since the relation between dose-rate and intensity is energy dependent, then different buildup factors will be appropriate to dose-rate and intensity calculations. For protection purposes the buildup factors appropriate to dose-rate are more useful, and therefore these alone are given in Table III.

TABLE III. DOSE BUILD-UP FACTORS FOR WATER, IRON AND LEAD
(AFTER GOLDSTEIN AND WILKINS[17])

| MeV | (a) Point isotropic geometry $\mu d$ | | | | | | MeV | (b) Plane monodirectional geometry $\mu d$ | | | | | |
|---|---|---|---|---|---|---|---|---|---|---|---|---|---|
| | 1 | 2 | 4 | 7 | 10 | 15 | 20 | | 1 | 2 | 4 | 7 | 10 | 15 |
| | *Water* | | | | | | | | *Water* | | | | | |
| 0.5 | 2.52 | 5.14 | 14.3 | 38.8 | 77.6 | 178.0 | 334.0 | 0.5 | 2.63 | 4.29 | 9.05 | 20.0 | 35.9 | 74.9 |
| 1 | 2.13 | 3.71 | 7.68 | 16.2 | 27.1 | 50.4 | 82.2 | 1 | 2.26 | 3.39 | 6.27 | 11.5 | 18.0 | 30.8 |
| 2 | 1.83 | 2.77 | 4.88 | 8.46 | 12.4 | 19.5 | 27.7 | 2 | 1.84 | 2.63 | 4.28 | 6.96 | 9.87 | 14.4 |
| 3 | 1.69 | 2.42 | 3.91 | 6.23 | 8.63 | 12.8 | 17.0 | 3 | 1.69 | 2.31 | 3.57 | 5.51 | 7.48 | 10.8 |
| 4 | 1.58 | 2.17 | 3.34 | 5.13 | 6.94 | 9.97 | 12.9 | 4 | 1.58 | 2.10 | 3.12 | 4.63 | 6.19 | 8.54 |
| 6 | 1.46 | 1.91 | 2.76 | 3.99 | 5.18 | 7.09 | 8.85 | 6 | 1.45 | 1.86 | 2.63 | 3.76 | 4.86 | 6.78 |
| 8 | 1.38 | 1.74 | 2.40 | 3.34 | 4.25 | 5.66 | 6.95 | 8 | 1.36 | 1.69 | 2.30 | 3.16 | 4.00 | 5.47 |
| 10 | 1.33 | 1.63 | 2.19 | 2.97 | 3.72 | 4.90 | 5.98 | | | | | | | |
| | *Iron* | | | | | | | | *Iron* | | | | | |
| 0.5 | 1.98 | 3.09 | 5.98 | 11.7 | 19.2 | 35.4 | 55.6 | 0.5 | 2.07 | 2.94 | 4.87 | 8.31 | 12.4 | 20.6 |
| 1 | 1.87 | 2.89 | 5.39 | 10.2 | 16.2 | 28.3 | 42.7 | 1 | 1.92 | 2.74 | 4.57 | 7.81 | 11.6 | 18.9 |
| 2 | 1.76 | 2.43 | 4.13 | 7.25 | 10.9 | 17.6 | 25.1 | 2 | 1.69 | 2.35 | 3.76 | 6.11 | 8.78 | 13.7 |
| 3 | 1.55 | 2.15 | 3.51 | 5.85 | 8.51 | 13.5 | 19.1 | 3 | 1.58 | 2.13 | 3.32 | 5.26 | 7.41 | 11.4 |
| 4 | 1.45 | 1.94 | 3.03 | 4.91 | 7.11 | 11.2 | 16.0 | 4 | 1.48 | 1.90 | 2.95 | 4.61 | 6.46 | 9.92 |
| 6 | 1.34 | 1.72 | 2.58 | 4.14 | 6.02 | 9.89 | 14.7 | 6 | 1.35 | 1.71 | 2.48 | 3.81 | 5.35 | 8.39 |
| 8 | 1.27 | 1.56 | 2.23 | 3.49 | 5.07 | 8.50 | 13.0 | 8 | 1.27 | 1.55 | 2.17 | 3.27 | 4.58 | 7.33 |
| 10 | 1.20 | 1.42 | 1.95 | 2.99 | 4.35 | 7.54 | 12.4 | 10 | 1.22 | 1.44 | 1.95 | 2.89 | 4.07 | 6.70 |
| | *Lead* | | | | | | | | *Lead* | | | | | |
| 0.5 | 1.24 | 1.42 | 1.69 | 2.00 | 2.27 | 2.65 | 2.73 | 0.5 | 1.24 | 1.39 | 1.63 | 1.87 | 2.08 | — |
| 1 | 1.37 | 1.69 | 2.26 | 3.02 | 3.74 | 4.81 | 5.86 | 1 | 1.38 | 1.68 | 2.18 | 2.80 | 3.40 | 4.20 |
| 2 | 1.39 | 1.76 | 2.51 | 3.66 | 4.84 | 6.87 | 9.00 | 2 | 1.40 | 1.76 | 2.41 | 3.36 | 4.35 | 5.94 |
| 3 | 1.34 | 1.68 | 2.43 | 3.75 | 5.30 | 8.44 | 12.3 | 3 | 1.36 | 1.71 | 2.42 | 3.55 | 4.82 | 7.18 |
| 4 | 1.27 | 1.56 | 2.25 | 3.61 | 5.44 | 9.80 | 16.3 | 4 | 1.28 | 1.56 | 2.18 | 3.29 | 4.69 | 7.70 |
| 6 | 1.18 | 1.40 | 1.97 | 3.34 | 5.69 | 13.8 | 32.7 | 6 | 1.19 | 1.40 | 1.87 | 2.97 | 4.69 | 9.53 |
| 8 | 1.14 | 1.30 | 1.74 | 2.89 | 5.07 | 14.1 | 44.6 | 8 | 1.14 | 1.30 | 1.69 | 2.61 | 4.18 | 9.08 |
| 10 | 1.11 | 1.23 | 1.58 | 2.52 | 4.34 | 12.5 | 39.2 | 10 | 1.11 | 1.24 | 1.54 | 2.27 | 3.54 | 7.70 |

When the incident beam is not monochromatic then the transmission through a barrier must be computed on the basis of the spectral distribution, bearing in mind that the narrow beam absorption coefficients and the broad beam buildup factors are both energy dependent. Spectral distributions of X-radiation produced by the interaction of monoenergetic electrons with various targets can be calculated, and worked examples have been published.[18] A complete calculation of the transmission of an X-ray beam through a thick absorber would therefore involve an integration of the product of the exponential absorption term and the buildup factor over the whole of the calculated spectrum. In practice approximate values of the transmission may be calculated by assuming that the incident beam behaves as a monochromatic beam having an energy one-third that of the peak energy.

Experimentally determined broad beam absorption curves for X-rays[19,20] are shown in Figs. 5 and 6.

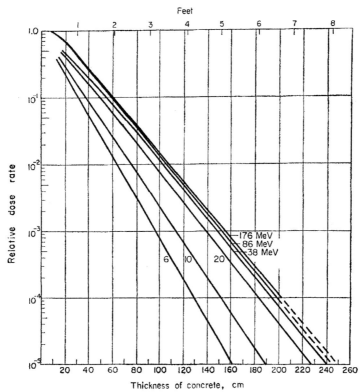

FIG. 5. Broad beam absorption in concrete of X-rays of various peak energies. (Concrete density 147 lb/ft³, 2.35 g/cm³.) After Kirn and Kennedy[19].

Shielding design should take account of primary radiation in all directions, not merely in the forward beam. In many instances off-axis radiation is removed by a collimator near the target. Shielding against scattered electromagnetic radiation at high energies is a minor problem. For high energy quanta the scattering absorption coefficients are small and the quanta are rapidly reduced in energy by multiple Compton scattering with resultant increase in absorption coefficient. As an example, the intensity of scattered radiation at a distance of 1 metre from 1 gram of water and at 45° to a primary beam of energy 15 and 22 MeV has been calculated to be about $3 \times 10^{-7}$ of that of the primary beam. This has been checked by direct observation.[21]

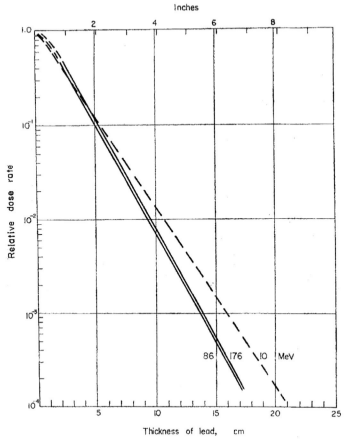

FIG. 6. Broad beam absorption in lead of X-rays of various peak energies.
(Lead density 11.35 g/cm³.)  After Miller and Kennedy[20].


APPENDIX III

PARTICLE RANGES

For estimating the shielding requirements for particles of different energies, the parameter of greatest usefulness is the practical range of the particles in such materials as water, air and lead. Figure 7, taken from Wachsmann and Dimotsis[22], shows the range of electrons, protons, deuterons and alpha particles in water, lead and air as a function of energy.

APPENDIX IV

ABSORPTION DATA FOR NEUTRONS

In many circumstances the specification of neutron shielding is difficult since the neutron flux density incident on the barrier can be only approximately predicted. Estimated neutron yields, energy distributions and spatial distributions should therefore be taken only as a guide, and should be checked by measurement on the particular installation. A bibliography[3] of neutron production data, including photoneutron yields, should be consulted.

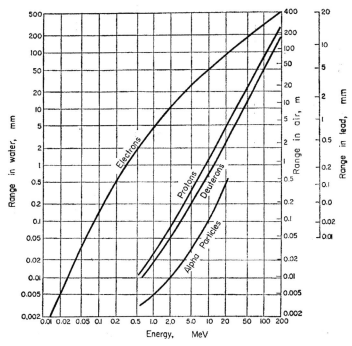

FIG. 7. Particle ranges in water, air and lead.
After Wachsmann and Dimotsis[22].

As with gamma-rays, the absorption of neutrons follows different laws for broad and narrow beams. For fast neutrons the absorption coefficient varies regularly with atomic weight. However, the absorption of thermal neutrons follows processes which do not vary smoothly with atomic weight. In any neutron shielding problem it is necessary to consider the production and subsequent absorption of gamma-radiation in the shielding. When the shield is subject to significant exposure to thermal neutrons the shield itself may become radioactive and so pose a further protection problem. It must be emphasized that neutron shielding calculations are not as straightforward as those for X-rays and that such calculations should always be based on a full understanding of the various processes.[23,24]

A narrow monoenergetic fast neutron beam is attenuated according to the formula :

$$I/I_0 = \exp\left(-\mu_t d\right)$$

where $I$ is the intensity at the point of measurement

$I_0$ is the intensity at the same point but in the absence of the absorber

$\mu_t$ is the macroscopic total cross-section for neutrons of the particular energy being considered

It is usual to express $\mu_t$ in cm²/g and to express $d$ in g/cm². Values of the macroscopic total cross-section for a particular material may be calculated from tabulated values of the microscopic (atomic) cross-sections of the element[25] by the following formulae.

$d$ is the barrier thickness measured in units reciprocal to those used for expressing the cross-section

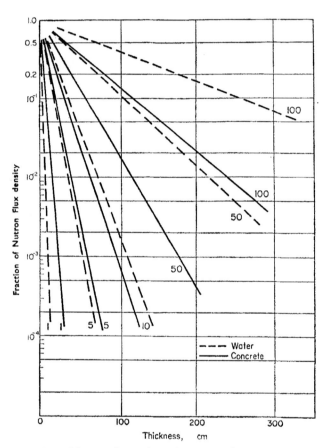

FIG. 8. Approximate broad beam absorption of neutrons in water and concrete (concrete density 148 lb/ft³, 2.37 g/cm³). Note that the curve shows the reduction of neutron flux density, not of dose rate.

For a pure material the macroscopic total cross-section is given by :

$$\mu_t = 0.602\,\sigma_t \rho/A \ (\text{cm}^{-1}) \qquad (2)$$

where    $\sigma_t$ = microscopic total cross-section for neutrons of the particular energy being considered (barns)

$\rho$ = the density of the material (g/cm³)

$A$ = the atomic weight of the material

For a material composed of several elements the cross-section is given by simple summation over its constituents :

$$\mu_t \text{ compound} = \mu_i'p' + \mu_i''p'' + \cdots \tag{3}$$

where     $\mu'_t, \mu''_t$, etc. are calculated from equation (2) for each element,
and        $p', p''$, etc. are the proportions (by weight) of each element in the compound.
  Worked examples of these formulae may be consulted[3].

Broad beams of fast neutrons are attenuated according to a law which is often sufficiently well represented by equation (1) in which $\mu_t$, the total cross-section, is replaced by $\mu_r$, the macroscopic removal cross-section, where $\mu_r = k\mu_t$. Between 1 MeV and 8 MeV it can be taken that $k = 0.7$. Above 8 MeV the value of $k$ decreases and experimental evidence[26] suggests that at 100 MeV it takes the value $k = 0.35$. Figure 8 shows absorption data calculated on the foregoing basis. The factor of proportionality between $\mu_r$ and $\mu_t$ at energies between 8 MeV and 100 MeV was estimated by linear extrapolation.

## REFERENCES

1. *Radiation Quantities and Units*, ICRU Report 10a (1962) National Bureau of Standards Handbook 84. Superintendent of Documents, U.S. Government Printing Office, Washington, D.C., U.S.A. (1962).
2. Recommendations of the International Commission on Radiological Protection (ICRP), *Brit. J. Radiol.* Suppl. 6 (1955).
3. *Protection Against Neutron Radiation up to 30 MeV*. Report of National Committee on Radiation Protection and Measurements, National Bureau of Standards Handbook 63. Superintendent of Documents, U.S. Government Printing Office, Washington, D.C., U.S.A. (1957).
4. P. BONET-MAURY, A. DEYSINE, M. FRILLEY and C. STEFAN, Efficacité biologique relative des protons de 157 MeV, *Comptes rendus* **251**, 3087 (1960).
5. B. LARSSON and B. A. KIHLMAN, Chromosome aberrations following irradiation with high energy protons and their secondary radiation : a study of dose distribution and biological efficiency using root-tips of *Vicia faba* and *Allium cepa*, *Int. J. Radiat. Biol.* **2**, 8 (1960).
6. G. J. NEARY and J. MULVEY, Maximum Permissible Fluxes of High Energy Neutrons and Protons in the Range 40–1000 MeV, Unpublished (revised by Neary, 1958).
7. *Report of the International Commission on Radiological Units and Measurements (ICRU)*, 1959, National Bureau of Standards Handbook 78. Superintendent of Documents, U.S. Government Printing Office, Washington, D.C., U.S.A. (1962).
8. *Measurement of Neutron Flux and Spectra for Physical and Biological Applications*. Recommendations of the National Committee on Radiation Protection and Measurements, National Bureau of Standards Handbook 72. Superintendent of Documents, U.S. Government Printing Office, Washington, D.C., U.S.A. (1960).
9. *Measurement of Absorbed Dose of Neutrons and of Mixtures of Neutrons and Gamma Rays*. Recommendations of the National Committee on Radiation Protection and Measurements, National Bureau of Standards Handbook 75. Superintendent of Documents, U.S. Government Printing Office, Washington, D.C., U.S.A. (1961).
10. *Health Physics in Nuclear Installations*. Proceedings of a Symposium organized by the European Nuclear Energy Agency at the Danish Atomic Energy Commission, Risø, 1959. Organization for European Economic Co-operation (1959).
11. *Selected Topics in Radiation Dosimetry*, Proceedings of a Symposium sponsored by the International Atomic Energy Agency, Vienna 1960. IAEA (1961).
12. P. HOWARD-FLANDERS, Physical and chemical mechanisms in the injury of cells by ionising radiations. *Advances in Biological and Medical Physics*, VI, Academic Press, New York (1958).
13. M. DANZKER, N. D. KESSARIS and J. S. LAUGHLIN, Absorbed dose and LET in radiation experiments, *Radiology* **72**, No. 1, 51 (1959).

14. L. V. SPENCER and U. FANO, Energy spectra resulting from electron slowing down, *Phys. Rev.* **93**, 1172 (1954).

15. L. V. SPENCER and F. H. ATTIX, A theory of cavity ionisation, *Radiation Research* **3**, 239 (1955).

16. G. W. GRODSTEIN, *X-ray Attenuation Coefficients from 10 keV to 100 MeV*. National Bureau of Standards Circular 583. Superintendent of Documents, U.S. Government Printing Office, Washington, D.C., U.S.A. (1957).

17. H. GOLDSTEIN and J. E. WILKINS, Calculations of the Penetration of Gamma Rays. USAEC Report, NYO 3075 (1954).

18. *Protection Against Betatron–Synchrotron Radiations up to 100 MeV*. National Bureau of Standards Handbook 55. Superintendent of Documents, U.S. Government Printing Office, Washington, D.C., U.S.A. (1954).

19. F. S. KIRN and R. J. KENNEDY, Betatron X-rays: How much concrete for shielding?, *Nucleonics* **12**, No. 6, 44–48 (1954).

20. W. MILLER and R. J. KENNEDY, Attenuation of 86 and 176 MeV synchrotron X-rays in concrete and lead, *Radiation Research* **4**, 360 (1956).

21. H. E. JOHNS, M. TUBIANA and J. L. HAYBITTLE, Unpublished.

22. F. WACHSMANN and A. DIMOTSIS, *Graphs and Tables for Radiotherapy*, Hirzel, Stuttgart (1957).

23. B. T. PRICE, C. C. HORTON and K. T. SPINNEY, *Radiation Shielding*, Pergamon Press, London (1957).

24. A. F. AVERY, B. E. BENDALL, J. BUTLER and K. T. SPINNEY, *Methods of Calculation for Use in the Design of Shields for Power Reactors*. AERE–R3216 (SWP/P52) HMSO, London (1960).

25. D. J. HUGHES and R. B. SCHWARTZ, *Neutron Cross Sections*. BNL 325, 2nd Edition. Superintendent of Documents, U.S. Government Printing Office, Washington, D.C., U.S.A. (1958) (see also BNL 325, 2nd Edition, Supplement No. 1 (1960); for angular distribution, see Hughes, D. J. and Carter, R. S. BNL 400 (1956)).

26. R. W. WILLIAMS, *Conference on Shielding of High Energy Accelerators*, USAEC Report TID 7545, Office of Technical Services, Department of Commerce, Washington, D.C., U.S.A. (1957).

# RECOMMENDATIONS OF THE INTERNATIONAL COMMISSION ON RADIOLOGICAL PROTECTION

## ORGANIZATION

THE International Commission on Radiological Protection originated in the Second International Congress of Radiology in 1928. Since then the Commission has had a close relationship with succeeding Congresses, and it has also been looked to as the appropriate body to give general guidance on the more widespread use of radiation sources caused by the rapid developments in the field of nuclear energy. The Commission wishes to maintain fully its traditional contact with medical radiology, and to fulfil its responsibilities to the medical profession. In addition, the Commission recognizes its responsibility to other professional groups and its obligation to provide guidance within the field of radiation protection as a whole.

The policy adopted by the Commission in preparing its recommendations is to deal with the basic principles of radiation protection, and to leave to the various national protection committees the responsibility of introducing the detailed technical regulations, recommendations, or codes of practice best suited to the needs of their individual countries.

The Commission has kept its recommendations continually under review in order to cover the increasing number and scope of potential radiation hazards, and to amend safety factors in the light of new knowledge concerning the effects of ionizing radiations.

Since the Ninth International Congress of Radiology in Munich the Commission has published two reports, entitled " Permissible Dose for Internal Radiation " (ICRP Publication 2), and " Protection Against X-rays up to Energies of 3 MeV and Beta- and Gamma-rays from Sealed Sources " (ICRP Publication 3). The latter report is essentially a code of practice for medical radiologists.

In 1959 the International Commission on Radiological Protection and the International Commission on Radiological Units and Measurements were jointly asked by the United Nations Scientific Committee on the Effects of Atomic Radiation (UNSCEAR) to prepare a report dealing with somatic effects of medical radiation exposures. A study group was appointed at the Munich meeting; it met in 1960, and prepared a report entitled " Exposure of Man to Ionizing Radiation Arising from Medical Procedures, with Special Reference to Radiation-Induced Diseases ". The report was submitted to the UNSCEAR at the end of 1960, and in 1961 it was published in *Physics in Medicine and Biology* **6**, No. 2 (Taylor & Francis Ltd., London).

Committee reports have also been prepared on :

(a) Protection Against Electromagnetic Radiation above 3 MeV and Electrons, Neutron and Protons;
(b) Handling and Disposal of Radioactive Materials in Hospitals and Small Establishments;
(c) RBE (Relative Biological Effectiveness).*

It is expected that these reports will be published during 1964.

The Commission has maintained close contact with the World Health Organization (WHO) and with the International Atomic Energy Agency (IAEA), with both of which the Commission has an official relationship. The Commission has been represented by observers at a number of meetings organized by the WHO and the IAEA. Co-operation has also been maintained with the International Labour Office (ILO), the Food and Agriculture Organization, and the UNSCEAR,

---

* Published in *Health Physics* **9**, No. 4 (1963).

all of which have been invited to send observers to technical meetings of the Commission and its committees. The Commission has been represented at all meetings of the UNSCEAR during the period 1960–1962.

During the period between the two last Congresses, the ILO has prepared an international instrument consisting of a Radiation Protection Convention (No. 115) supplemented by a Recommendation (No. 114). The Commission realizes that the reference to its work, in paragraphs 3, 4 and 5 of the Recommendation, increases its responsibility and adds to the importance of keeping its recommendations continually under review.

Grants of money have been made to the Commission by a number of organizations. The Ford Foundation has allocated $250,000, to be paid over a period of 5 years. The World Health Organization contributed $9,000 in the years 1960 and 1961, and $10,000 in 1962 and 1963. The International Society of Radiology gave $3,000 for the period between the 1959 and 1962 Congresses and $3,000 for 1963. A sum of $10,000 was received from the UNSCEAR in connection with the work of the Joint Study Group described above. The International Atomic Energy Agency has contributed $6,000 for 1963. The Commission wishes to express its sincere appreciation to all these organizations.

A meeting of the Commission and its committees was held in Stockholm in May 1962. The Commission also met in Ottawa in executive sessions immediately before the Tenth International Congress of Radiology; at these meetings the Commission prepared the present report incorporating its latest recommendations as well as amendments to previous recommendations.

Since the meeting in Munich the Commission has suffered the death of two of its most distinguished members; its Chairman Emeritus, Sir Ernest Rock Carling, and its Vice-Chairman Emeritus, Professor G. Failla. The Commission owes a great debt of gratitude to them for their invaluable work during the most significant years in the history of radiation protection.

During the preparation of this statement the ICRP has had the following composition:

### 1959–1962

R. M. SIEVERT, *Chairman* (Sweden)
E. E. POCHIN, *Vice-Chairman* (Great Britain)
W. BINKS (Great Britain)
L. BUGNARD (France)
H. HOLTHUSEN (Germany)
J. C. JACOBSEN (Denmark)
R. G. JAEGER (Germany)

J. F. LOUTIT (Great Britain)
K. Z. MORGAN (U.S.A.)
H. J. MULLER (U.S.A.)
R. S. STONE (U.S.A.)
L. S. TAYLOR (U.S.A.)
E. A. WATKINSON (Canada)

SIR ERNEST ROCK CARLING, *Chairman Emeritus* (Great Britain)—Deceased 1960
G. FAILLA, *Vice-Chairman Emeritus* (U.S.A.)—Deceased 1961

B. LINDELL, *Secretary* (Sweden)

### 1962

E. E. POCHIN, *Chairman* (Great Britain)
L. BUGNARD, *Vice-Chairman* (France)
W. BINKS (Great Britain)
O. HUG (Germany)
H. JAMMET (France)
B. LINDELL (Sweden)
J. F. LOUTIT (Great Britain)

K. Z. MORGAN (U.S.A.)
H. J. MULLER (U.S.A.)
R. M. SIEVERT (Sweden)
C. G. STEWART (Canada)
R. S. STONE (U.S.A.)
L. S. TAYLOR (U.S.A.)

F. D. SOWBY, *Scientific Secretary* (Canada)

During the Stockholm meeting in May, 1962, the Commission decided to reorganize the structure of its committees. The previous committees were dissolved at the time of the Tenth International Congress of Radiology and four new committees were established to review various topics of interest to the Commission :

1. Radiation Effects
2. Internal Exposure
3. External Exposure
4. Application of Recommendations.

The following have accepted invitations to serve as members of these committees :

### Committee 1

J. F. LOUTIT, *Chairman* (Great Britain)
F. DEVIK (Norway)
A. R. GOPAL-AYENGAR (India)
O. HUG (Germany)
L. F. LAMERTON (Great Britain)

J. LEJEUNE (France)
H. B. NEWCOMBE (Canada)
R. SCOTT RUSSELL (Great Britain)
A. C. UPTON (U.S.A.)

### Committee 2

K. Z. MORGAN, *Chairman* (U.S.A.)
W. BINKS (Great Britain)
A. M. BRUES (U.S.A.)
B. CHR. CHRISTENSEN (Denmark)
M. IZAWA (Japan)
M. LAFUMA (France)

L. D. MARINELLI (U.S.A.)
W. G. MARLEY (Great Britain)
E. E. POCHIN (Great Britain)
V. SHAMOV (U.S.S.R.)
W. S. SNYDER (U.S.A.)
C. G. STEWART (Canada)

### Committee 3

E. E. SMITH, *Chairman* (Great Britain)
J. DUTREIX (France)
R. G. JAEGER (Austria)
L.-E. LARSSON (Sweden)

A. PERUSSIA (Italy)
E. DALE TROUT (U.S.A.)
B. M. WHEATLEY (Great Britain)
H. O. WYCKOFF (U.S.A.)

### Committee 4

H. JAMMET, *Chairman* (France)
D. J. BENISON (Argentina)
G. C. BUTLER (Canada)
H. DAW (U.A.R.)
H. J. DUNSTER (Great Britain)
B. LINDELL (Sweden)

D. MECHALI (France)
C. POLVANI (Italy)
P. RECHT (Belgium)
C. P. STRAUB (U.S.A.)
E. G. STRUXNESS (U.S.A.)
F. WESTERN (U.S.A.)

This reorganization does not change, in essence, the scope of program of the previous Committees I and II, but the old Committees III, IV and V and the ICRP/ICRU Committee on RBE are not included as such under the new structure. All problems relating to external exposure, both for quantum and for particulate radiation at any energy, will now be considered within the new Committee 3. Special *ad hoc* task groups will be set up from time to time to deal with specific problems.

The Committees I–V and the *ad hoc* Committee on RBE had the following composition during the period 1959–1962.

### Committee I (*Advisory Committee on Biology*)

J. F. LOUTIT, *Chairman* (Great Britain)
A. A. BUZZATI-TRAVERSO (Italy)
J. A. COHEN (The Netherlands)
F. DEVIK (Norway)
A. R. GOPAL-AYENGAR (India)
L. F. LAMERTON (Great Britain)
J. LEJEUNE (France)

H. B. NEWCOMBE (Canada)
J. NIELSEN (Denmark)
R. SCOTT RUSSELL (Great Britain)
A. C. STEVENSON (Great Britain)
A. C. UPTON (U.S.A.)
S. WARREN (U.S.A.)

*Committee II* (*Protection against radiation from internal radioactive substances*)

K. Z. MORGAN, *Chairman* (U.S.A.)
W. BINKS (Great Britain)
A. M. BRUES (U.S.A.)
B. CHR. CHRISTENSEN (Denmark)
M. IZAWA (Japan)
W. H. LANGHAM (U.S.A.)

L. D. MARINELLI (U.S.A.)
W. G. MARLEY (Great Britain)
M. POBEDINSKI (U.S.S.R.)
E. E. POCHIN (Great Britain)
W. S. SNYDER (U.S.A.)
C. G. STEWART (Canada)

*Committee III* (*Protection against X-rays and electrons up to energies of 3 MeV and beta- and gamma-rays from sealed sources*)

R. G. JAEGER, *Chairman* (Germany)
E. E. SMITH, *Vice-Chairman* (Great Britain)
C. B. BRAESTRUP (U.S.A.)
E. D. TROUT (U.S.A.)
C. GARRETT (Canada)
F. GAUWERKY (Germany)
H. HOLTHUSEN (Germany)

L. LORENTZON (Sweden)
S. B. OSBORN (Great Britain)
C. POLVANI (Italy)
D. J. STEVENS (Australia)
H. O. WYCKOFF (U.S.A.)
J. ZAKOVSKY (Austria)
A. ZUPPINGER (Switzerland)

*Committee IV* (*Protection against electrons and electromagnetic radiation above 3 MeV, neutrons and radiation from heavy particle accelerators*)

G. J. NEARY, *Chairman* (Great Britain)
J. W. BOAG (Great Britain)
F. HERČÍK (Czechoslovakia)
G. S. HURST (U.S.A.)
H. E. JOHNS (Canada)
G. JOYET (Switzerland)

W. H. KOCH (U.S.A.)
J. S. LAUGHLIN (U.S.A.)
C. A. TOBIAS (U.S.A.)
M. TUBIANA (France)
B. M. WHEATLEY (Great Britain)
K. G. ZIMMER (Germany)

*Committee V* (*Handling of radioactive isotopes and disposal of radioactive waste*)

C. P. STRAUB, *Chairman* (U.S.A.)
L. R. DONALDSON (U.S.A.)
H. J. DUNSTER (Great Britain)
H. JAMMET (France)
A. W. KENNY (Great Britain)
C. A. MAWSON (Canada)

A. A. PERUSSIA (Italy)
E. H. QUIMBY (U.S.A.)
F. D. SOWBY (Canada)
E. G. STRUXNESS (U.S.A.)
F. WESTERN (U.S.A.)

*ICRP/ICRU Committee on RBE* (*Relative biological effectiveness*)

L. F. LAMERTON, *Chairman* (Great Britain)
J. F. LOUTIT (Great Britain)
H. H. ROSSI (U.S.A.)
W. S. SNYDER (U.S.A.)
G. J. NEARY (Great Britain)

O. HUG (Germany)
H. I. KOHN (U.S.A.)
H. QUASTLER (U.S.A.)
M. TUBIANA (France)
A. C. UPTON (U.S.A.)

## RULES GOVERNING THE SELECTION AND WORK OF THE INTERNATIONAL COMMISSION ON RADIOLOGICAL PROTECTION

1. (a) The International Commission on Radiological Protection (ICRP) shall be composed of a Chairman and not more than twelve other members. The selection of the members shall be made by the ICRP from nominations submitted to it by the National Delegations to the International Congress of Radiology and by the ICRP itself. The selections shall be subject to approval by the International Executive Committee (IEC) of the Congress. Members of the ICRP shall be chosen on the basis of their recognized activity in the fields of medical radiology, radiation protection, physics, health physics, biology, genetics, biochemistry and biophysics, with regard to an appropriate balance of expertise rather than to nationality.

(b) The membership of the ICRP shall be approved during each International Congress for service until the end of the succeeding Congress, or until new members are appointed. Not less than two but not more than four members shall be changed at any one Congress. In the intervening period vacancies may be filled by the ICRP.

(c) In the event of a member of the ICRP being unable to attend the ICRP meetings, a substitute may be selected by the ICRP as a temporary replacement. Such a substitute shall not have voting privileges unless specifically authorized by the ICRP.

(d) The ICRP shall be permitted to invite individuals to attend its meetings to give special technical advice. Such persons shall not have voting privileges, but their opinions may be recorded in the minutes.

2. The Chairman shall be elected by the ICRP from among its members to serve until the end of the succeeding Congress, or until his successor is elected. The choice shall not be limited to the country in which it is proposed to hold the succeeding Congress. The Chairman shall be responsible for reporting the proceedings and recommendations of the ICRP at the next Congress.

3. The ICRP shall elect from among its members a Vice-Chairman who will serve in the capacity of Chairman in the event that the Chairman is unable to perform his duties.

4. Minutes of meetings and records of the ICRP shall be made by a Scientific Secretary selected by the Chairman of the ICRP, subject to the approval of its members. The Scientific Secretary need not be a member of the ICRP. The records of the ICRP shall be passed on to the succeeding Scientific Secretary.

5. The Chairman, in consultation with the Vice-Chairman and the Scientific Secretary, shall prepare a program to be submitted to the Commission for discussion at its meetings. Proposals to be considered shall be submitted to the Chairman for circulation to all members of the ICRP and other specially qualified individuals at least two months before any meeting of the ICRP.

6. Decisions of the ICRP shall be made by a majority vote of the members. A minority opinion may be appended to the minutes of a meeting if so desired by any member, upon his submission of the same in writing to the Scientific Secretary.

7. The ICRP may establish such committees as it deems necessary to perform its functions.

# RECOMMENDATIONS OF THE COMMISSION

## A. PREFATORY REVIEW*

(1) Prior to the Geneva meeting of the Commission in April 1956, permissible levels of exposure to ionizing radiation had been expressed in terms of a dose in a rather short interval of time (1 day or 1 week), that is, essentially, in terms of an average dose-rate—the average referring to the temporal distribution of the dose in the specified interval of time. Implicitly, if not explicitly, it was assumed that if this average dose-rate was low enough, no appreciable bodily injury would become apparent in the lifetime of the individual. The assumption was based largely on radiological experience which indicated that substantial skin recovery occurred within a few months following a moderate therapeutic dose and that the latent period for some long-term effects of radiation (e.g. cancer of the skin) resulting from residual tissue damage, was longer the lower the dose (or dose-rate in the case of chronic exposure). Thus, in an occupationally exposed individual a long-term effect might not become apparent in his lifetime, even if a certain amount of permanent injury had occurred.

(2) The basic permissible weekly dose at that time was 0.3 rem/week. Assuming that a person was occupationally exposed at this rate (50 weeks a year) for 50 years, the permissible accumulated dose would be 750 rems in the most critical organs or essentially throughout the body. It was realized then that this constituted a " large " lifetime dose and an appropriate warning was included in the Commission's report of 1955.

(3) The general awareness of radiation hazards induced caution on the part of those responsible for the protection of workers. Administratively, liberal factors of safety were often used especially in large atomic energy installations. As a result it was found that in general the actual exposure of personnel was

kept at levels much below the then existing permissible limits.

(4) At the 1956 meeting of the Commission it became evident that stricter recommendations were needed. The 1955 Conference on the Peaceful Uses of Atomic Energy had aroused great interest in the development of atomic power plants throughout the world. In time this would greatly increase the number of persons occupationally exposed and would also bring about actual or potential exposure of other persons and the population as a whole. More importantly, the pressure for producing power economically might well do away with the " safety factors " mentioned above. Also, the average duration of occupational exposure per individual worker might increase. On the biological side it was considered that perhaps " recovery " plays a less important part in the long-term effects of radiation to be expected from continued exposure at low levels, than was earlier supposed. Because of the larger number of persons who would be exposed, occupationally or otherwise, genetic damage assumed greater importance. This was accentuated in no small degree by the realization that in some countries the per capita genetic dose contributed by medical procedures was about the same as that contributed by background radiation.

(5) Statistical studies had shown that the incidence of leukemia in radiologists was significantly greater than in other physicians who presumably were not professionally exposed to radiation. Of necessity these radiologists included those who had practiced their speciality at the time when radiation protection was not very effectively carried out. Therefore, the accumulated doses received by those who developed leukemia may have been much higher than the 750 rems mentioned above. On the other hand, since most of the exposure of these radiologists resulted from diagnostic

---

* Reproduced from the Commission's 1958 Recommendations. Footnotes added in 1962.

procedures carried out with low voltage X-rays, the lifetime dose in the blood-forming organs may have been lower than 750 rems even if the skin dose, especially in some parts of the body, was much higher. The mechanism of leukemia induction by radiation is not known. It may be postulated that if the dose is lower than a certain threshold value no leukemia is produced. In this case it would be necessary to estimate the threshold dose and to make allowances for recovery, if any. There is not sufficient information to do this, but caution would suggest that an accumulated dose of 750 rems might exceed the threshold. The most conservative approach would be to assume that there is no threshold and no recovery, in which case even low accumulated doses would induce leukemia in some susceptible individuals, and the incidence might be proportional to the accumulated dose. The same situation exists with respect to the induction of bone tumors by bone-seeking radioactive substances.

(6) Presently available longevity studies differ as to whether there is a statistically significant life shortening in radiologists as compared to other specialists presumably not occupationally exposed to radiation. However, in mammals chronically exposed at different daily doses a definite effect on longevity becomes clearly apparent at the higher daily doses. If extrapolation to lower daily doses, and then to man, is justified, it may be concluded that occupational exposure at presently accepted permissible limits may entail some life shortening. This effect may be interpreted as a slight acceleration of the natural aging process.

(7) The effects just discussed illustrate the two different types of possible long-term somatic effect that must be considered in setting up permissible limits of exposure. The first type (leukemia) is a serious effect occurring in some individuals and, therefore, the aim of protection would be to reduce the incidence to the lowest practical limit. The second type (life shortening) is presumably an effect on every individual and, therefore, the aim of protection would be to reduce the degree of effect to the lowest practical value. The definition of

permissible dose has been changed to include explicitly these two types of possible effect.

(8) Genetic effects manifest themselves in the descendants of exposed individuals. The injury, when it appears, may be of any degree of severity from inconspicuous to lethal. A slight injury will tend to occur in the descendants for many generations, whereas a severe injury will be eliminated rapidly through the early death of the individual carrying the defective gene. Thus the sum total of the effect caused by a defective gene until it is eliminated may be considered to be roughly the same. The main consideration in the control of genetic damage (apart from aspects of individual misfortune) is the burden to society in future generations imposed by an increase in the proportion of individuals with deleterious mutations. From this point of view it is immaterial in the long run whether the defective genes are introduced into the general pool by a few individuals who have received large doses of radiation, or by many individuals in whom smaller doses have produced correspondingly fewer mutations. However, even in this case it is desirable to limit the dose received by an individual.

(9) In view of the foregoing, recommendations are made in this report in terms of maximum permissible doses for individuals and for population groups. In either case limits are set on the basis of dose accumulated over a period of years rather than in terms of a weekly dose that could be received for an indefinite period of time. The concept of limiting the accumulated dose was introduced by the Commission at its 1956 meeting in Geneva. The limitation of accumulated dose suggested at the time corresponds roughly to a three-fold reduction in weekly dose, for example, in the case of whole body occupational exposure when the exposure takes place approximately at a constant rate.

(10) In practice the problem of chief concern is chronic exposure either at low dose rates or by intermittent small doses. Under these conditions it is reasonable to assume that the dose accumulated over a period of years is the controlling factor, *provided* the intermittent

C

doses are sufficiently small. Thus, in addition to limiting the accumulated dose it is necessary to limit the magnitude of a single dose (that is, a dose received in a short interval of time). Previously a single exposure equal to the maximum permissible weekly dose (" seven consecutive days ") was permitted. Following the same pattern, the single dose limit for occupational exposure recommended in the present report is expressed in terms of the maximum permissible dose accumulated in a period of " 13 consecutive weeks ". The recommended value for the relevant organ (e.g. 3 rems for the blood-forming organs) has been made as high as it appears prudent, in the light of present knowledge. The stipulation of any 13 *consecutive* weeks has been made to make sure that operations are carried out in such a way that intermittent doses approximating the full 13 week quota do not occur at short intervals.

(11) In the recommendations published in 1955 maximum permissible limits were set on the basis of doses received by certain organs and certain serious late effects known to occur in them with sufficiently large doses. Provisions were made by means of an arbitrary " dose distribution curve " (in the report of Committee I) to limit the dose in other organs and tissues. This was made necessary by the adoption of a maximum permissible dose for the skin twice as large as that for the blood-forming organs (with an assumed effective depth of 5 cm). In the present report separate recommendations are made for three groups of organs or tissues :

(a) Blood-forming organs, gonads and lenses of the eyes.*
(b) Skin and thyroid gland.
(c) All other organs or tissues, specifically as regards exposure essentially limited to the organ or tissue in question.

(12) For the blood-forming organs, gonads and the lenses of the eyes* the limits for occupational exposure are set in terms of the dose accumulated at various ages, according to the formula $D = 5(N - 18)$, where $D$ is the dose in rems and $N$ is the age in years, with the additional stipulation that the dose accumulated during any 13 consecutive weeks shall not exceed 3 rems.

(13) For the skin and the thyroid gland the limit for occupational exposure is set in terms of the dose accumulated during any 13 consecutive weeks, and the recommended value is 8 rems. This is derived from an average of 0.6 rem/week (the maximum permissible weekly dose formerly recommended for the skin of the whole body) which in 13 weeks amounts to 7.8 rems, and the nearest whole number is used to avoid the implication of greater accuracy than is warranted by present knowledge. The limit for the dose in these tissues accumulated in 1 year is $(0.6 \times 50) = 30$ rems. It should be noted that the new recommendation refers to the dose in the skin itself, irrespective of the dose distribution in the subcutaneous tissues. Therefore, the comparison should be made with the previous recommendation for exposure to radiation of very low penetrating power, for which the recommended limit was 1.5 rem/week. Accordingly, in this case also a reduction has been made in the accumulated dose, but the single exposure limit has been increased from 1.5 to 8 rems. This should provide more flexibility in practice than was possible formerly.

(14) For all organs and tissues of the body except the blood-forming organs, the gonads and the lenses of the eyes,* the limit for occupational exposure is set in terms of the dose accumulated during any 13 consecutive weeks. With the exception of the skin, the pertinent practical cases in this category relate to exposure from internal sources essentially limited to individual organs or tissues. The following points require consideration. Whereas in the case of the blood-forming organs, the gonads, the lenses of the eyes* and the skin, the objective of protection is to prevent or minimize definitely known types of injury, in the case of other organs the type of injury is not known. (Bone constitutes the only exception, in which case the relevant injury is cancer and permissible limits may be set on the basis of data furnished

---

* See footnote to paragraph 27.

by human subjects who accumulated radium in their skeletons.) Possibly, radiation in sufficient dosage may increase the incidence of cancer in one of these organs (e.g. the thyroid gland) or it may accelerate aging of the organ. In the absence of factual data, it was deemed prudent in earlier recommendations of the Commission to set the maximum permissible limit for these organs, when irradiated by internal sources, as low as that for the more sensitive organs such as the gonads, that is, 0.3 rem/week. When the exposure is essentially limited to *one organ* because of the more or less selective accumulation of a certain radioactive isotope therein, it is obvious that this limit embodies a factor of safety not present when the *whole body* is exposed at the same permissible limit. For this reason and the fact that none of these organs is known to be as sensitive as the blood-forming organs, the gonads and the lenses of the eyes,* the Commission has decided to retain the previously recommended maximum permissible dose of 0.3 rem/week for each organ singly (with some exceptions noted in the report of Committee II). However the limit is now expressed in terms of 13 consecutive weeks, which makes it 4 rems, in round figures, with an annual accumulated dose of 15 rems. Committee II has made the necessary adjustments to conform with the lower permissible limits now recommended for some organs and for what may be regarded as constituting " whole body " exposure (e.g. isotopes distributed throughout the body, or several isotopes present simultaneously, each concentrating significantly in a different organ).

(15) The Commission has given particular attention to the difficult problem of setting permissible limits for exposure of persons in the neighborhood of radiation installations. The chief obstacle is the almost complete lack of knowledge of the deleterious effects that may result from low level exposure starting at conception and continuing throughout life. It is reasonable to expect a more marked effect then in the case of exposure starting after the individual has reached maturity (for one thing,

because the period of exposure is longer), but it is very difficult to decide what allowance to make. Guidance could be obtained from suitable experiments carried out with mammals and it is hoped that such studies will be undertaken soon in some laboratories. In the meantime caution is in order. The Commission recommends that provisions be made in a controlled area or areas to make sure that under normal operating conditions no child residing outside such controlled areas, receive more than 0.5 rem/year (in the appropriate organs) from radiation or radioactive material originating in the controlled area or areas. In practice it may be expected that while fluctuations in exposure rate would occur they would not be such as to require special limitations. It will be noted that this is one tenth of the *lowest* annual dose in any organ permitted for occupational exposure. It includes contributions made by external and internal sources but does not include doses contributed by natural background radiation or medical procedures.

(16) Special groups of *adults* in the vicinity of a controlled area are permitted to receive larger annual doses in the gonads, the blood forming organs and the lenses of the eyes, by factor of three (i.e. 1.5 rems). No biological significance should be attached to the magnitude of this factor, since present radiobiological information is grossly inadequate in this respect. The value recommended (1.5 rems/year) is one-tenth of the former maximum permissible annual dose for occupational exposure, on the basis of 0.3 rem/week in the most sensitive organs.*

(17) Planning for the future expansion of nuclear energy programs and the more extensive uses of radiation, requires measures intended to protect whole populations. Genetic damage is of greatest concern in this regard. The problem has been discussed by various national and international groups and tentative suggestions have been made. The Commission considered the problem at its 1956 meeting and later issued a statement in general terms

---

* See footnote to paragraph 27.

* See also paragraphs 54 and 55.

However, recommendations in quantitative terms are needed in the design of power plants and other radiation installations and particularly in making plans for disposal of radioactive waste products. It is of the utmost importance in this connection to make sure that nothing is done now that may prove to be a serious hazard later, which cannot be corrected at all or will be very expensive to correct. The Commission is aware of the fact that a proper balance between risks and benefits cannot yet be made, since it requires a more quantitative appraisal of both the probable biological damage and the probable benefits than is presently possible. Furthermore, it must be realized that the factors influencing the balancing of risks and benefits will vary from country to country and that the final decision rests with each country (insofar as operations within one country do not affect other countries).

(18) The Commission wishes to point out that it is important to assign quotas of a maximum permissible genetic dose to the different modes of exposure, in order to make sure that those responsible for the control of exposure in one category do not take up a disproportionate share of the permissible total in their planning. However, at this time it is deemed best not to assign rigid quotas. As a tentative guide an illustrative apportionment is appended to paragraph 65.

(19) Briefly, the suggested limit for the genetic dose was arrived at in the following manner : Estimates made by different national and international scientific bodies indicate that a per capita gonad dose of 6–10 rems accumulated from conception to age 30 from all man-made sources, would impose a considerable burden on society due to genetic damage, but that this additional burden may be regarded as tolerable and justifiable in view of the benefits that may be expected to accrue from the expansion of the practical applications of " atomic energy ". There is at present considerable uncertainty as to the magnitude of the burden (see for example the report of the United Nations Scientific Committee on the Effects of Atomic Radiation) and, therefore, it is highly desirable to keep the exposure of large populations at as low a

level as practicable, with due regard to the necessity of providing additional sources of energy to meet the demands of modern society. A genetic dose of 10 rems from all man-made sources is regarded by most geneticists as the absolute maximum and all would prefer a lower dose. In some countries the genetic dose from medical procedures has been estimated to be about 4.5 rems* (see *Report of Joint Study Made by ICRP/ICRU for the U.N. Scientific Committee*). Therefore, if the limit for the genetic dose from all man-made sources were set at 6 rems, the contribution from all sources, other than medical procedures, would be limited to 1.5 rems in these countries. This would impose unacceptable restrictions on these countries. Accordingly, as a matter of practical necessity the Commission recommends that medical exposure be considered separately and that it be kept as low as is consistent with the necessary requirements of modern medical practice. The joint study of ICRP/ICRU indicates that careful attention to the protection of the gonads would result in a considerable reduction of the genetic dose due to medical procedures without impairment of their value. In view of these considerations the Commission suggests a limit of 5 rems for the genetic dose from all man-made sources of radiation and activities, except medical procedures.

(20) At the present time the contribution to the genetic dose from all man-made sources (other than medical procedures) is small. With careful planning the rate of increase can be kept under control and the ultimate value of this contribution may never reach the suggested limit of 5 rems. Since the genetic dose from medical exposure in most countries is much lower than 4.5 rems and since in those countries in which it is high efforts are being made to reduce it, the total genetic dose from all man-made sources actually received by the world population may be expected to be considerably less than 10 rems, perhaps even less than 6 rems in the foreseeable future. Furthermore, if a thermonuclear reaction can be utilized as a source of power, the problem

---

* See also paragraph 70.

of radiation protection may be greatly simplified.

(21) The Commission is aware that compliance with the new recommendations may entail structural changes in some existing installations and/or changes in operative procedures. Since in fact the new recommendations are more restrictive because of the greater emphasis put on the dose accumulated over a long period of time, it is not essential that such changes be made immediately, although it is obviously desirable. As a practical guide it is suggested that the transition period during which the necessary changes would be made, should not exceed 5 years.*

(22) The Commission wishes to point ou again that the setting up of maximum per missible limits of occupational and non occupational exposure (especially the latter requires quantitative information not yet avail able about the risks and benefits of an expandec use of " atomic energy ". For this reason th Commission will be glad to receive factual data and suggestions from those concerned with th production or utilization of ionizing radiation so that as much pertinent information a possible may be available to it in its futur deliberations.

## B. BASIC CONCEPTS†

### OBJECTIVES OF RADIATION PROTECTION

(23) Exposure to ionizing radiation can result in injuries that manifest themselves in the exposed individual and in his descendants : these are called somatic and genetic injuries respectively.

(24) Late somatic injuries include leukemia and other malignant diseases, impaired fertility, cataracts and shortening of life. Genetic injuries manifest themselves in the offspring of irradiated individuals, and may not be apparent for many generations. Their detrimental effect can spread throughout a population by mating of exposed individuals with other members of the population.

(25) The objectives of radiation protection are to prevent or minimize somatic injuries and to minimize the deterioration of the genetic constitution of the population.

### CRITICAL ORGANS AND TISSUES

(26) The organs and tissues of the body exhibit varying degrees of radiosensitivity, and it is therefore necessary, for purposes of protection, to consider their radiosensitivity with

respect to specific functions as well as the dose they receive. When this is done, some organ and tissues assume a greater importance according to the circumstances under whicl they are irradiated. They are then said to b critical.

(27) In the case of more or less uniforn irradiation of the *whole body*, the critical tissue are those tissues of the body that are mos radiosensitive with respect to the ability o carrying out functions essential to the body as a whole. In this report these are taken to be th blood-forming organs, the gonads, and the lense of the eyes.‡ In previous reports the skir was listed as a critical organ in the case o whole body exposure. The presentation of the recommendations in the present report is simplified by not designating the skin as a critical organ.

(28) In the case of irradiation more or less limited to *portions* of the body, the critical tissue is that tissue most likely to be permanently damaged either because of its inherent radiosensitivity, or because of a combination of radiosensitivity and localized high dose.

---

* It will be noted that this suggestion was made in 1958.

† Revised in 1962.

‡ In the 1958 Recommendations the blood-forming organs, the gonads and the lenses of the eyes were regarded as critical tissues in the case of whole body exposure. There is evidence indicating that the lens may be specially radiosensitive only to particulate radiation of high LET (e.g. neutrons having an energy of 1 MeV). On the evidence available at present the lens seems not to assume greater importance than other tissues when X-, gamma- and beta-radiations only are concerned (see paragraph 52c).

### SIGNIFICANT AREAS AND VOLUMES

(28a) The Commission's 1955 recommendations (*Brit. J. Radiology*, Supplement 6) included a reference to the volume or area over which the dose should be averaged in the computation of maximum permissible tissue doses. This reference has not been republished in subsequent recommendations but it is recommended that the following principles should apply.

(28b) Within permissible dose ranges specified for occupational exposure, when the object of control is to reduce to a very low order of magnitude the risk of late effects (such as malignant change from an accumulated dose of radiation) it is justifiable to consider the average dose to the whole organ or tissue. This has practical advantages in that the significant volume can be taken as that of the organ or tissue under consideration. In fact, this principle has necessarily been used already in calculating permissible concentrations of radioactive nuclides in tissues.

(28c) When skin is contaminated with radioactive material, it is considered that the previous recommendation of a significant area of 1 cm² might be too restrictive. It is therefore recommended that the significant area in such cases be taken to be of the order of 30 cm². This is a practicable recommendation from the standpoint of procedures used to determine the degree of contamination of the skin.

(28d) In other cases of external exposure, especially when the distance to the source is very short or when the exposed area is very small, it would not be appropriate to recommend the dose to be assessed as the average over 30 cm²; for these cases it is recommended that the previous practice of referring to 1 cm² be maintained.

### PERMISSIBLE DOSE

(29) Any departure from the environmental conditions in which man has evolved may entail a risk of deleterious effects. It is therefore assumed that long continued exposure to ionizing radiation additional to that due to natural radiation involves some risk. However, man cannot entirely dispense with the use of ionizing radiations, and therefore the problem in practice is to limit the radiation dose to that which involves a risk that is not unacceptable to the individual and to the population at large. This is called a " permissible dose ".

(30) The permissible dose for an *individual* is that dose, accumulated over a long period of time or resulting from a single exposure, which, in the light of present knowledge, carries a negligible probability of severe somatic or genetic injuries; furthermore, it is such a dose that any effects that ensue more frequently are limited to those of a minor nature that would not be considered unacceptable by the exposed individual and by competent medical authorities.

(31) Any severe somatic injuries (e.g. leukemia) that might result from exposure of individuals to the permissible dose would be limited to an exceedingly small fraction of the exposed group; effects such as shortening of life span, which might be expected to occur more frequently, would be very slight and would likely be hidden by normal biological variations. The permissible doses can therefore be expected to produce effects that could be detectable only by statistical methods applied to large groups.

(32) The permissible dose to the gonads for the *whole population* is limited primarily by considerations with respect to genetic effects (see paragraphs 58–65).

(32a) On the basis of the criteria indicated in paragraphs 29–32, the Commission has given recommendations with regard to the *maximum* dose which, still fulfilling the above requirements, should be permitted under various circumstances. The Commission has balanced as far as possible the risk of the exposure against the benefit of the practice, and has also considered the possible danger involved in remedial actions once the exposure has occurred. This dose has been called the *maximum permissible dose.*

(32b) The basis of the Commission's recommendations is that any exposure to radiation may carry some risk. The assumption has been made that, down to the lowest levels of dose, the risk of inducing disease or disability in an individual increases with the dose accumulated by the individual, but is small even at the maximum permissible levels recommended for occupational exposure. This assumption is supported by the limited statistics available which indicate that for radiation workers of the last generation, exposed subject to the maximum permissible levels of that time, the risks of somatic effects are comparable with or less than those of the majority of other trades and professions, and would therefore be considered as not unacceptable. The Commission similarly considered the risk of somatic effects in individuals within certain population groups, and recommended maximum permissible levels for these individuals.

(32c) With regard to genetic effects, the Commission assumed that the genetic burden to a population will be proportional to the genetic dose received by that population (see paragraph 63 of the 1958 Recommendations). The Commission therefore recommended a maximum permissible genetic dose of 5 rems,* on the basis that the resulting burden to society would be " tolerable and justifiable in view of the benefits that may be expected to accrue from the expansion of the practical application of 'atomic energy' ".

(32d) The implication of a maximum permissible dose is that it must be capable of being controlled. It is therefore clear that the Commission's recommended maximum permissible doses are appropriate for those conditions in which the levels can be controlled. However, in the case of accidents and of environmental contamination when exposures may not be subject to control, the concept of a fixed maximum permissible dose ceases to be meaningful. Instead, other considerations arise, such as the need to balance the risk from radiation against the risks of particular countermeasures. The principles upon which these risks might be assessed and balanced are still under consideration.

### DOSE-RATE EFFECTS

(32e) *Somatic effects.* It has long been recognized that the effects of radiation may be

---

\* From all sources additional to natural background plus the lowest practicable contribution from medical exposure.

dependent not only on the accumulated dose received but also on the way in which this total is fractionated and on the dose-rate at which each fraction is given. This applies particularly to radiation of low LET such as X-rays, gamma-rays and beta-radiation which are by far the commonest radiations encountered in occupational practice at the present time. It appears possible on some theoretical and experimental grounds that when either the total dose or the dose-rate is very low and effects will be directly proportional to the total dose and independent of dose-rate. This assumption is implicit in past recommendations on permissible levels and although confirmatory proof is lacking it is believed to be a reasonable basis for assessment.

(32f) *Genetic effects.* A linear dose-effect relationship unaffected by dose-rate has been generally accepted in the past for gene mutations. Recent experimental work has shown, however, that at intermediate and higher levels of dose-rate the number of mutations produced in test-subjects may not be independent of dose-rate. Because of the importance of the genetic effect the Commission has made a special survey of recent work. This survey indicates that evidence for dependence of mutation frequency on dose-rate comes at present almost entirely from a study of several gene-loci in spermatogonial cells and oocytes of the mouse. For insects there is not yet clear evidence of a difference in effectiveness of acute and chronic doses that cannot be ascribed to selection. No general relationship appears therefore to hold for all species.

(32g) On the basis of the above, the Commission does not at present modify its recommendations to allow for dose-rate effects in man.

### CATEGORIES OF EXPOSURE

(33) These recommendations are designed to limit not only somatic but also genetic effects; it is therefore necessary to reduce as much as possible the dose to the population as a whole, as well as to the individual. In general, doses resulting from all sources of ionizing radiation should be considered in the appraisal of possible biological damage. How

ver, practical considerations make it necessary to consider separately the doses resulting from two categories of exposure, namely:

    (a) Exposure to natural background radiation.

    (b) Exposure resulting from medical procedures.

(34) Natural background radiation varies considerably from locality to locality and the doses it contributes to the various organs are not well known. If maximum permissible limits recommended by the Commission included background radiation, the allowable contribution from man-made sources—which are the only ones that can be controlled—would be uncertain and would have to be different for different localities. Accordingly, doses resulting from natural background radiation are excluded from all maximum permissible doses recommended in this report.

(35) In medical procedures, exposure of the patient to primary radiation is generally limited to parts of the body, but the whole body is exposed to some extent to stray radiation. The contributions to the doses in various organs and the part played in the over-all effects on the individual are practically impossible to evaluate at the present time. The Commission recognizes especially the importance of the *gonad* doses resulting from medical exposure and the attendant genetic hazard to the population. Accordingly, it recommends that the medical profession exercise great care in the use of ionizing radiation in order that the gonad dose received by individuals before the end of their reproductive periods be kept at the minimum value consistent with medical requirements. However, individual doses resulting from medical exposure are excluded from all maximum permissible doses recommended in this report.

(36) The recommendations cover the following three categories* of exposed individuals:

---

\* It should be noted that exposures within the first two categories as defined here correspond to the two classes of exposure used in the International Labour Office's Radiation Protection Convention, namely (1) Workers directly engaged in radiation work, and (2) Workers not directly engaged in radiation work.

The first category consists of individuals who are occupationally exposed to radiation.

The second category comprises adults who work in the vicinity of controlled areas or who enter controlled areas occasionally in the course of their duties, but who are not themselves employed on work involving exposure to radiation.

The third category consists of individual members of the population at large (including persons living in the neighborhood of controlled areas).

In addition to recommending maximum permissible doses for individuals, the Commission also gives separate recommendations about the average exposure to the population as a whole, determined by exposures of the individuals within the above categories (paragraphs 58–70c).

*Occupational Exposure*

(37) Exposure of an *individual* who normally works in a controlled area constitutes occupational exposure. Maximum permissible doses are set for the individuals in the small portion of the population that can be occupationally exposed (paragraphs 46–52). The contribution from this group to the genetic dose to the *population* as a whole is discussed in paragraph 65.

*Exposure of Adult Workers not Directly Engaged in Radiation Work*

(38) Adults who work in the vicinity of controlled areas or who enter controlled areas occasionally in the course of their duties may be exposed to radiation originating in the controlled area, even if they are not themselves employed on work which ordinarily involves exposure to radiation. This category may include women of reproductive age and individuals subject to other hazards, and for these reasons the maximum permissible dose to these individuals is set lower than for persons occupationally exposed (paragraphs 53–55). The contribution from this category to the genetic dose to the whole population is discussed in paragraph 65.

*Exposure of Individual Members of the Population at Large*

(39) Individual members of the population at

large may be exposed to radiation originating in a controlled area or to radiation that cannot be related to any specific controlled area; e.g. exposure from environmental contamination and widely distributed radiation sources such as wrist-watches, TV-sets and various applications of radioactive materials to be expected as a result of future expansion in the atomic energy field. As such exposure is not easily controlled, it will be impossible to ensure that a recommended maximum permissible individual dose is not exceeded in any single case. Where large numbers are involved, it will not be possible to examine the habits of every individual. A reasonable procedure would be to study a sample of the group involved and to set the environmental level so that no individual in the sample receives any excessive exposure. There will always remain the possibility that someone of grossly different habits from those in the observed sample may receive a higher dose than the maximum in the sample.

(40) In order to facilitate planning for the anticipated increased uses of nuclear energy and other sources of radiation, it is desirable at this time to recommend a maximum for the genetic dose to the *population* (paragraph 64); this maximum will determine what average gonad exposure could be allowed. Part of the recommended maximum genetic dose will have to be used for occupational exposure, for exposure of adult workers not directly engaged in radiation work, and for medical exposure. The proper apportionment for exposure of the population at large must allow for both internal and external exposure (paragraph 65).

*Medical Exposure*

(41) No recommendations are given with regard to the dose to the individual from medical exposure. (The contribution of medical exposure to the genetic dose is discussed ir paragraphs 69–70c.)

### REDUCTION IN MAXIMUM PERMISSIBLE DOSE

(42) The 1958 recommendations were introduced partly with the intention of limiting the genetically significant radiation exposure (see paragraph 63) of the population, and partly to limit the probability of somatic injury by reducing the lifetime dose. This reduction does not result from positive evidence of damage due to the use of the earlier permissible dose levels, but rather is based on the concept that biological recovery may be minimal at such low dose levels.

### TIME INTERVAL OVER WHICH DOSE IS TO BE ASSESSED

(43) The maximum permissible weekly dose recommended by the Commission in 1950 were replaced in 1958 by limits for the doses received over longer periods of time (paragraphs 47–49) In the case of occupational exposure the maximum permissible dose that may be accumulated at a certain time depends on the age of the worker. The dose to individuals in the population at large, or in special group other than occupational, may be accumulated at a rate that is determined by a maximum permissible annual dose. The genetic dose to the whole population is assessed over the period between conception of the individual and conception of each child of the individual (See paragraph 63 for method of evaluation.)

(44) These extended periods of time allow for some flexibility in the way in which radiation exposure may be received, and at the same time provide what is considered to be adequate protection for each group of the population.

## C. MAXIMUM PERMISSIBLE DOSES*

### GENERAL

(45) It is emphasized that the maximum permissible doses recommended in this section are *maximum* values; the Commission recommends that all doses be kept as low as practicable, and that any unnecessary exposure be avoided.

---

* Revised in 1962.

## EXPOSURE OF INDIVIDUALS

### OCCUPATIONAL EXPOSURE

(46) In any organ or tissue, the *total* dose due to occupational exposure shall comprise the dose contributed by external sources during working hours and the dose contributed by internal sources taken into the body during working hours. It shall not include any medical exposure or exposure to natural radiation. The Commission wishes to emphasize that " medical exposure " refers to the exposure of *patients* that is necessary for medical purposes and *not* to the exposure of the personnel conducting such procedure.

### Exposure of the Gonads and the Blood-forming Organs

(47) The maximum permissible total dose accumulated in the gonads and the blood-forming organs at any age over 18 years shall be governed by the relation

$$D = 5(N - 18)$$

where $D$ is tissue dose in rems* and $N$ is age in years.

(48) For a person who is occupationally exposed at a constant rate from age 18 years, the formula implies a maximum weekly dose of 0.1 rem. It is recommended that this value be used for purposes of planning and design.

### Rate of Dose Accumulation

(49) To the extent the formula permits, an occupationally exposed person may accumulate the maximum permissible dose at a rate not in excess of 3 rems during any period of 13 consecutive weeks (i.e. in no 13 consecutive weeks shall the dose exceed 3 rems).† If necessary, the 3 rems may be received as a single dose, but as the scientific knowledge of the biological effects of differing dose-rates is scant, single doses of the order of 3 rems should be avoided as far as practicable.

(49a) *Exposure of women of reproductive age.* The recommendation for dose accumulation at rates up to 3 rems per quarter should not apply in circumstances involving abdominal exposure of women of reproductive age. Women of reproductive age should be occupationally employed only under conditions where the exposure of the abdomen is limited to 1.3 rems in a 13 week period, corresponding to 5 rems per year delivered at an even rate. Under these conditions the dose to an embryo during the first two months of pregnancy would normally be less than 1 rem, a dose which the Commission considers to be acceptable.

(49b) *Exposure of pregnant women.* When a pregnancy has been diagnosed, arrangements should be made to ensure that the exposure of the woman be such that the average dose to her fetus during the remaining period of the pregnancy does not exceed 1 rem. Under conditions in which the fetal dose may approximate to that received by the woman, for example when the abdomen is exposed to penetrating radiation, this recommendation will normally be met if the woman is not exposed at rates greater than those applicable to adult workers not directly engaged in radiation work, since such workers may not be exposed at rates greater than 1.5 rems per year (see paragraph 54).

(49c) Under conditions in which the fetal dose will be considerably less than that received by the woman, as for example when working with diagnostic X-ray equipment of 150 kV or less, or where the abdomen is protected from or not exposed to the radiation, the recommendation may be satisfied by continuing occupational exposure of the woman at a rate not exceeding 1.3 rems per 13 weeks.

### Application to Special Cases*

(50) Setting permissible limits of exposure in terms of the dose accumulated up to a given age introduces certain practical complications. Thus, some workers (previously exposed at levels within the then permissible limits) may have already accumulated a dose in excess of the maximum permitted by the formula. There

---

* See earlier discussion of RBE and $QF$.

† A calendar 13 week period starting at any chosen date may be used instead of a period of 13 consecutive weeks if there is no reason to suppose that doses are being accumulated at grossly irregular rates.

---

* Paragraphs 50 to 51e are reproduced from the 1958 Recommendations of the Commission.

are also special cases in which exceptions in the application of the formula may be desirable for practical reasons and are justifiable within the context of paragraph 42. The following recommendations are intended to provide guidance on administrative policy, which may well vary according to circumstances at the local level. (It should be noted that this situation will obtain only during a relatively short transition period.)

(51a) *Previous exposure history unknown.* When the previous occupational exposure history of an individual is not definitely known, it shall be assumed that he has already received the full quota permitted by the formula.

(51b) *Persons exposed in accordance with the former maximum permissible weekly dose.* Persons who were exposed in accordance with the former maximum permissible weekly dose of 0.3 rem and who have accumulated a dose higher than that permitted by the formula, should not be exposed at a rate higher than 5 rems in any one year, until the accumulated dose at a subsequent time is lower than that permitted by the formula.

(51c) *Persons starting work at an age of less than 18 years.* When a person begins to be occupationally exposed at an age of less than 18 years, the dose shall not exceed 5 rems in any one year under age 18, and the dose accumulated to age 30 shall not exceed 60 rems. (The minimum age at which occupational exposure is legally permitted is lower than 18 years in some countries.)

(51d) *Accidental high exposure.* An accidental high exposure that occurs *only once in a lifetime* and contributes no more than 25 rems shall be added to the occupational dose accumulated up to the time of the accident. If the sum then exceeds the maximum value permitted by the formula, the excess need not be included in future calculations of the person's accumulated dose. Accidental exposure to doses higher than 25 rems must be regarded as being potentially serious, and shall be referred to competent medical authorities for appropriate remedial action and recommendations on subsequent occupational exposure. This is intended as an administrative guide to permit the continuation

of work with radiation, following a bona fide accident (" once in a lifetime "), in cases in which interruption of such work, or curtailment of exposure, would handicap the individual in the pursuit of his career.

(51e) *Emergency exposure.* Emergency work involving exposure above permissible limit shall be planned on the basis that the individual will not receive a dose in excess of 12 rems. This shall be added to the occupational dose accumulated up to the time of the emergency exposure. If the sum then exceeds the maximum value permitted by the formula, the excess shall be made up by lowering the subsequent exposure rate so that within a period not exceeding 5 years, the accumulated dose will conform with the limit set by the formula. Women of reproductive age shall not be subjected to such emergency exposure.

*Exposure of Single Organs other than the Gonads and the Blood-forming Organs*

(52) For exposure that is essentially restricted to portions or single organs of the body, with the exception of the gonads and the blood-forming organs, a higher dose than the one derived from the formula $D = 5(N - 18)$ is permitted. The following recommendations are made.

(52a) *A maximum dose of 8 rems/13 weeks for the skin, thyroid and bone.** The maximum dose in the skin, thyroid and bone, accumulated over any 13 consecutive weeks, shall not exceed 8 rems. This recommendation applies to all exposure of the skin, except the skin of the hands and forearms, feet and ankles (see paragraph 52b). As the 8 rems are derived from an average of 0.6 rem/week, the annual dose for a 50 week year is limited to 30 rems.

(52b) *A maximum dose of 20 rems/13 weeks for the hands and forearms, feet and ankles.* In the case of exposure of the hands and forearms, feet and ankles, the maximum dose accumulated over any 13 consecutive weeks shall not exceed 20 rems. This recommendation applies to all tissues of the above-mentioned extremities. As the 20 rems are derived from an average of

---

* The dose to bone is based on a body-burden of 0.1 $\mu$c of Ra$^{226}$ (see Report of Committee II, ICRP Publication 2).

.5 rems/week, the annual dose for a 50 week year is limited to 75 rems.

(52c) *A maximum dose of 4 rems/13 weeks for limited exposure of internal organs\* other than the thyroid, the gonads and the blood-forming organs.* In the case of internal organs, limited exposure originates almost exclusively from radioisotopes within the body. As most planning of release of radioactive isotopes to the air and water supplies in controlled areas is made under conditions where occupational groups rather than single individuals are exposed, a formula corresponding to the one given in paragraph 47 cannot in general be applied to internal exposure. An average of 0.3 rem/week in the organ of interest (with some exceptions noted in the report of Committee II) is expected to be maintained under equilibrium conditions if the concentration in air *or* water of the relevant isotope is kept at levels given in the tables in the report of Committee II. Variations of the dose-rate will occur in practice, and are permissible, provided that the dose accumulated over any 13 consecutive weeks does not exceed 4 rems. As this maximum is derived from a weekly average of 0.3 rem, the annual dose for a 50 week year is limited to 15 rems. In the case of irradiation of the lens of the eye with particulate radiation of high LET (e.g. neutrons having an energy of 1 MeV) a special $QF$ of 30 shall be used, instead of the usual value of 10.

(52d) *Whole body exposure from uptake of several radioisotopes.* When the radioactive isotopes in a mixture are taken up by several organs and the resultant tissue doses in such organs are of comparable magnitude, the combined exposure is considered to constitute essentially whole body exposure. Accordingly, the permissible levels of exposure will be those applicable to the gonads and the blood-forming organs.

(52e) *Short-term exposures to radioactive materials.* One or more short-term exposures to radioactive materials within a period of 13 consecutive weeks is considered acceptable if the total *intake* of radioactive material during this period does not exceed the cumulative intake allowed when exposure occurs for 13 weeks at the maximum levels (MPC values) for occupational exposure given in the Report of Committee II (ICRP Publication 2). If significant exposure to external sources occurs concurrently, the quarterly intake referred to above should be estimated to make allowance for the dose equivalent delivered by external sources (ICRP Publication 2, page 24). The 50 year integrated dose to the critical organ from such an intake will not exceed 1.3 rems for the whole body, blood-forming organs or gonads, 8 rems for skin, thyroid and bone,\* and 4 rems for other organs.

(52f) *Chemical toxicity.* Because of the toxicity of natural uranium, $U^{238}$, $U^{236}$, or $U^{235}$ in soluble form in amounts permitted according to the above on the basis of radiological protection, the inhalation of uranium of any isotopic composition should not exceed 2.5 milligrams of soluble uranium in one day, or the ingestion averaged over 2 days should not exceed 150 milligrams of soluble uranium.† For insoluble forms of uranium, maximum intake shall be limited by the same rule that applies to other radioactive materials set forth above. Because of the low specific activity of certain other radioactive materials with very long radioactive half-lives (e.g. $Rb^{87}$, $In^{115}$, $Nd^{144}$, $Sm^{147}$, $Re^{187}$, etc.), the mass or chemical toxicity will in general determine maximum rates of intake. Similar considerations will apply also to the values given in Table 1 of ICRP Publication 2.

(52g) *Accidental high exposure to radioactive materials.* In the case of an accidental high exposure to radioactive materials where the total intake of radioactive material exceeds the amount that would result from intake for 1 year at the maximum levels for occupational exposure

---

\* The dose to bone is based on a body-burden of 0.1 $\mu$c of $Ra^{226}$ (see Report of Committee II, ICRP Publication 2).

† This restriction on the ingestion of soluble forms of uranium revises the maximum permissible concentration recommended in Table 1 of ICRP Publication 2 for the continuous ingestion of drinking water. The new values of MPC for uranium that result from taking into account the chemical toxicity of uranium are given in Part 5 of ICRP Publication 6.

---

\* Including the lens of the eye.

to such radioactive materials given in the Report of Committee II (ICRP Publication 2), an estimate of the intake resulting from the exposure shall be entered on the individual's record and he shall be referred to competent medical authorities for appropriate action.

(52h) *Emergency work involving exposure to radioactive materials.* Emergency work involving exposure to radioactive materials at levels above the normal maximum permissible concentrations shall be planned on the basis that the total intake of radioactive material during the emergency period should not exceed the cumulative intake that would result from exposure for 1 year at the maximum levels (MPC values) for occupational exposure to such radioactive materials given in the Report of Committee II (ICRP Publication 2). If significant exposure to external sources might be expected to occur concurrently, the annual intake referred to above should be estimated to make allowance for the dose equivalent delivered by the external sources (ICRP Publication 2, page 24). The 50 year integrated dose to the critical organ from such an intake will not exceed 5 rems for the whole body, blood-forming organs or gonads, 30 rems for skin, thyroid and bone,* and 15 rems for other organs. When such an exposure has occurred, an estimate of the intake shall be entered in the individual's record, and action shall be taken appropriate to paragraphs 86 and 86a.

### ADULT WORKERS NOT DIRECTLY ENGAGED IN RADIATION WORK

(53) In any organ or tissue, the total maximum permissible individual dose shall consist of the sum of the doses contributed by both external and internal sources. It shall not include any medical exposure or exposure to natural radiation.

(54) For adult workers not directly engaged in radiation work (see paragraph 36) the total annual dose, including contributions from external and internal sources, to the gonads and the blood-forming organs, shall not exceed 1.5 rems, nor shall the contribution from a mixture of isotopes whose combined exposure constitutes essentially whole body exposure make the total annual dose exceed 1.5 rems.

(55) The annual dose to all other organs and tissues, and to the hands and forearms, feet and ankles, shall be limited to one-tenth of the corresponding annual occupational doses.

### INDIVIDUAL MEMBERS OF THE POPULATION AT LARGE

(56) In any organ or tissue the total maximum permissible individual dose shall consist of the sum of the doses contributed by both external and internal sources. It shall not include any medical exposure or exposure to natural radiation.

(57) The population at large contains children, for whom it is considered that a lower dose to the gonads and the blood-forming organs should apply. The maximum total dose limit for individuals in the population at large shall be 0.5 rem per year to the gonads and the blood-forming organs, nor shall the contribution from a mixture of isotopes whose combined exposure constitutes essentially whole body exposure make the total annual dose exceed 0.5 rem.

(57a) The annual dose to all other organs and tissues and to the hands and forearms, feet and ankles, shall be limited to one-tenth of the corresponding annual occupational doses. This recommendation applies to exposures for which control can be exercised. In the case of accident and of environmental radiation beyond regular control, special considerations arise (see the above discussion on permissible dose).

## EXPOSURE OF POPULATION†

### GENERAL

(58) Proper planning for nuclear power programs and other peaceful uses of atomic energy on a large scale requires a limitation of the exposure of whole populations, partly by limiting the individual doses and partly by limiting the number of persons exposed.

(59) This limitation necessarily involves a

---

* The dose to bone is based on a body-burden of 0.1 $\mu c$ of $Ra^{226}$ (see Report of Committee II, ICRP Publication 2).

† This section deals with exposures averaged over whole populations.

ompromise between deleterious effects and social benefits. Consideration of genetic effects plays a major role in its evaluation. The problem has been discussed extensively in recent years and suggestions have been made by different national bodies. The Commission is aware of the fact that a proper balance between risks and benefits cannot yet be made, since it requires a more quantitative appraisal of the probable biological damage and the probable benefits than is presently possible. Furthermore, it must be realized that the factors influencing the balancing of risks and benefits will vary from country to country and that the final decision rests with each country.

(60) Because of the urgent need for guidance in this regard, the Commission in its 1958 Recommendations suggested an interim ceiling for the exposure of the whole population. The proposed level is essentially in accordance with suggestions by other scientific groups that have studied the problem and tried to evaluate the possible biological effects. It is felt that this level provides reasonable latitude for the expansion of atomic energy programs in the foreseeable future. It should be emphasized that the limit may not in fact represent the proper balance between possible harm and probable benefit, for reasons already mentioned.

(61) On the assumption that the genetic effects are linearly related to the gonad dose and provided that no threshold dose exists, it is possible to define a population dose average that is relevant to the assessment of genetic injury to the whole population (paragraphs 62-63). In the case of somatic effects no such dose can easily be defined although the annual per capita dose to certain tissues or to the whole body may be relevant on the assumption of a non-threshold, linear dose-effect relation.

### GENETIC DOSE

*Assessment of Genetic Dose*

(62) The genetic dose to a population is the dose which, if it were received by each person from conception to the mean age of childbearing, would result in the same genetic burden to the whole population as do the actual doses received by the individuals. A permissible genetic dose is that dose, which if it were received by each person from conception to the mean age of childbearing, would result in an acceptable burden to the whole population.

(63) The genetic dose to a population can be assessed as the annual genetically significant dose multiplied by the mean age of childbearing, which for the purpose of these recommendations is taken to be 30 years. The annual genetically significant dose to a population is the average of the individual gonad doses, each weighted for the expected number of children conceived subsequent to the exposure.

*Maximum Permissible Genetic Dose*

(64) It is suggested that the genetic dose (see paragraph 63) to the whole population from all sources additional to the natural background should not exceed 5 rems plus the lowest practicable contribution from medical exposure. The background is excluded from the suggested value because it varies considerably from country to country. The contribution from medical exposure is considered separately for the same reason and also because the subject is being studied for the purpose of limiting such exposure to the minimum value consistent with medical requirements.

*Apportionment of Genetic Dose*

(65) The suggested maximum genetic dose of 5 rems in addition to the dose from medical procedures and natural background must not be used up by one single type of exposure. The proper apportionment of the total will depend upon circumstances which may vary from country to country, and the decision should therefore be made by national authorities.

*Addendum to Paragraph 65*

(a) *Illustrative apportionment.* The Commission does not wish to make any firm recommendations as to the apportionment of the genetic dose of 5 rems but, for guidance, gives the following apportionment as an illustration :

| | |
|---|---|
| Occupational exposure .. .. .. | 1.0 rem |
| Exposure of adult workers not directly engaged in radiation work .. | 0.5 rem |

Exposure of the population at large .. 2.0 rem
Reserve .. .. .. .. .. 1.5 rem

(b) *Fractions of population.* Assuming that the ratio of the total population to the reproductive population is the same in all groups, the largest fraction ($\varepsilon$) of the whole population that can be exposed to an average annual dose $\overline{D}_1^i$ is given by the equation:

$$\varepsilon.30.\overline{D}_1^i = D_{30}^i$$

where $D_{30}^i$ is the apportionment of the genetic dose to the $i$th exposure group, and the average annual dose within the group can be expressed as a fraction of the maximum permissible individual annual dose; i.e. $\overline{D}_1^i = F_1 D_1^i$.

(c) *Occupational exposure.* Assigning 1.0 rem to occupational exposure would mean that 0.7 per cent of the whole population could accumulate the maximum permissible occupational gonad dose of 60 rems by age 30. It is very unlikely that such a figure will be approached in the foreseeable future. At the present time, the number of persons occupationally exposed in technologically developed countries is about 0.1–0.2 per cent of the population, and most of these persons receive doses which are considerably less than the maximum permissible doses.

(d) *Exposure of adult workers not directly engaged in radiation work.* An apportionment of 0.5 rem for the exposure of adult workers not directly engaged in radiation work would imply that about 1 per cent of the population could receive the maximum permissible individual annual gonad doses for these individuals. The allowable size of this group varies inversely with the average dose within the group. Thus, if this dose were only 10 per cent of the maximum individual doses, the group could amount to 10 per cent of the whole population, which is larger than is likely to occur.

(e) *Exposure of the population at large.* The apportionment of 2.0 rems (with a long-term reserve of 1.5 rems for possible eventualities) for the genetic exposure of the population at large is intended for planning purposes in the development of nuclear energy programs (with the associated waste disposal problems) and more extensive uses of radiation sources. In the case of internal exposure, the radioisotopes of interest are those that contribute to the gonad dose directly (by local concentration) or indirectly (by radiation produced elsewhere in the body). In either case the maximum permissible concentrations in air and water of these isotope recommended by Committee II for continuou occupational exposure (" 168 hour week ") ar based on an average yearly dose of 5 rems in th gonads or the whole body. If for these isotopes th average concentrations in public air and wate supplies are lower than the values recommende for continuous occupational exposure by a facto of 1/100, the genetic dose to the population woul amount to 1.5 rems (5 × 1/100 rems/year in th gonads × 30 years = 1.5 rems). In this case th contribution from external sources should b limited to 0.5 rem in order not to exceed th total of 2 rems.

## SOMATIC DOSE

(66) No specific recommendations are mad at this time as to the maximum permissibl " somatically " relevant dose to the population However, it is expected that the maximun permissible limits of the *individual* total dose recommended in paragraphs 46–57a will kee the average dose in any tissue at such a leve that the injuries that could possibly occur in population would be well within acceptabl limits (see paragraph 31).

(67) In the case of external exposure of th whole body to penetrating radiation th restriction imposed by the genetic dose to th population automatically reduces the doses t internal organs below the *individual* maximun permissible annual doses recommended i paragraphs 46–57a. Thus the probability o somatic injury in these organs is considerabl lower than indicated in paragraph 66. Th same thing applies to internal exposure resultin from radioisotopes that directly or indirectl contribute to the gonad dose of a populatio (see addendum to paragraph 65).

(68) There remain for further consideratio those isotopes that concentrate in specifi organs (other than the gonads). In view of th existing uncertainty as to the dose-effec relationships for somatic effects, it is suggeste that for planning purposes the average con centrations of such isotopes, or mixture thereof, in air or water, applicable to th population at large, should not exceed one thirtieth of the MPC values for continuou

occupational exposure given in the report of Committee II.*

### MEDICAL EXPOSURE

*Radiological Examinations of Women of Reproductive Age*

(69) The Commission wishes to call attention to reports of embryonic and fetal sensitivity to ionizing radiation and to emphasize that the possibility of pregnancy must be taken into account by the attending physician when deciding on radiological examinations that involve the lower abdomen and pelvis of women of reproductive age. The Commission also wishes to point out that the 10-day interval following the onset of menstruation is the only time when it is virtually certain that women of such age are not pregnant. Therefore, it is recommended that all radiological examinations of the lower abdomen and pelvis of women of reproductive age, that are not of importance in connection with the immediate illness of the patient, be limited in time to this period when pregnancy is improbable. The examinations that should be delayed to await the onset of the next menstruation are those that could without detriment be delayed till the conclusion of a pregnancy or at least until its latter half.

*Apportionment of Total Genetic Dose—Medical Exposure*

(70a) In the 1958 Recommendations (para-graph 70) the Commission referred to the upper limit of the estimates of the annual genetically significant dose from medical exposure, and noted that the highest levels could be reduced significantly by careful attention to techniques.

(70b) More recent estimates of the genetic dose from medical exposure, described in the 1962 Report of UNSCEAR,† indicate that the actual dose may be considerably less than the highest levels referred to above. In addition, the 1962 UNSCEAR Report referred to estimates, which have been made in two countries, of the extent by which a reduction in dose could be achieved without loss of necessary medical information.‡

(70c) Paragraph 70 of the 1958 Recom-mendations concluded: " A certain allowance for medical exposure of populations must be made and may be made in the near future as the subject is being studied for the purpose of limiting such exposure to the minimum value consistent with medical requirements ". It is clear that the permissible dose will depend upon the value of the medical benefit to be derived, and future extensions in the use of radiological methods may confer greater benefit from their application than detriment from the necessary associated exposure to radiation. At the present time the Commission has decided to maintain its policy of not making numerical recom-mendations with regard to the appropriate genetic dose from medical exposures.

## D.  GENERAL PRINCIPLES REGARDING WORKING CONDITIONS§

### RESPONSIBILITY

(70d) The owner or the person in charge of controlled area shall be responsible for the working conditions and for the instruction of all persons working in the area regarding radiation hazards and methods of control. He shall be responsible for maintaining radiation levels outside the controlled area resulting from

---

* The basis for the limits of permissible exposure of populations to man-made sources of ionizing radiations the dose received by the various organs of the body and not the MPC values, or other criteria by which the dose is controlled. Nevertheless, for planning purposes some guidance as given in paragraph 68 must be available. The word average " in paragraph 68 refers to the concentration of radioactive nuclides, averaged over a year, in the total intake to the average person of the population.

† Estimates of the annual genetically significant dose from X-ray diagnosis in 14 countries ranged from to 60 mrem.

‡ Ministry of Health, Department of Health for Scotland, Radiological Hazards to Patients. Second report of the Committee. HMSO (1960).

LARSSON, L.-E. *Acta Radiologica*, Supplement 157 : 7–127 (1958).

§ Reproduced from the Commission's 1958 Recommendations with the addition of paragraphs 84a and 86a.

operations within the area so that exposures do not exceed the maximum permissible doses recommended in paragraphs 53–57a.

(71) A controlled area shall be established where persons occupationally exposed could receive doses in excess of 1.5 rems/year.

(72) A controlled area is an area in which the exposure of personnel to radiation or radioactive material is under the supervision of a radiation protection officer.

(73) A radiation protection officer is one who has the knowledge and responsibility to apply appropriate radiation protection regulations. He may be the owner or the person in charge of the controlled area or he may be a technically competent person appointed by the above.

(74) A qualified expert (or health physicist) is a person having the knowledge and training needed to measure ionizing radiations and to advise regarding radiation protection. The qualification should be of the type specified by a National Committee.

## RADIATION SURVEYS AND MONITORING

### Surveys Prior to Use of Controlled Areas

(76) In those instances where the operations in a controlled area may disturb or alter significantly the environment with respect to radiation hazards, adequate surveys should be made of the radioactivity of the air, soil and water prior to the start of operations. This will provide a base line from which to judge the adequacy of radiation controls within the area.

(77) During and after installation, appropriate radiation surveys shall be made to ensure that the pertinent recommendations will be complied with when routine operations commence. Routine operations shall be deferred until such compliance is assured.

(78) When additional operations are planned in the area, a thorough survey should be made of the background radiation prior to the start of the new operations. This will aid in the identification of the operation responsible for any increase of the background radiation or the contamination in the area.

### Routine Surveys and Monitoring

(79) Radiation surveys shall be made regularly, at a frequency dictated by the operations within the area, to determine the adequacy of safety procedures. This should include checks of the facilities, equipment (radiation warning devices, radiation shields, hoods, respirators ventilating system, etc.), and working techniques. When there is any reasonable probability of a radiation hazard existing, the vicinity of the controlled area should also be surveyed.

### Special Surveys

(80) Specific and detailed recommendation regarding radiation surveys applicable to some special cases are given in the Reports of the Committees of ICRP.

## HEALTH SURVEILLANCE

### Pre-employment Examination

(81) All new personnel in radiation work shall have a pre-employment medical examination. Notes should be made of the family history, of the previous occupational history, and of previous X-ray diagnostic examination or radiation therapy. The examination shall include a complete blood count, with determination of erythrocyte and leukocyte levels and a differential white cell count. It should be recognized that the examination is directed toward determining the " normal " condition of the worker at the time of employment, and toward noting any abnormalities that might later be confused with radiation damage.

(82) In cases where there has been previous occupational exposure, the total accumulated dose shall be recorded (see paragraphs 47–51 and any appropriate additional medical examinations performed. These should include ophthalmological examinations, with particular reference to changes in the lens, in cases of exposure to neutrons and to heavy particles and examinations of skin and nails in the case of partial external irradiation and external contamination.

*Routine Medical Examinations*

(83) Medical examinations should be performed at a frequency depending upon the conditions of the occupational exposure. Blood counts, although they are a part of a medical examination, are not to be considered as a method of radiation monitoring.

(84) Persons occupationally exposed to neutrons and to heavy particles of significant penetration should have ophthalmological examinations, with particular reference to changes in the lens. The frequency of the examination will depend upon the conditions of exposure.

*Occupational Working Hours and Length of Vacation*

(84a) The Commission considers that with the present maximum permissible exposure levels no special treatment of radiation workers with respect to working hours and length of vacation is needed.

### PERSONNEL MONITORING

*External Radiation*

(85) Doses received as a result of occupational exposures shall be systematically checked with appropriate instruments to ensure that the maximum permissible doses are not exceeded and to make it possible to keep individual cumulative dose records.

*Internal Radiation*

(86) Tests should be performed to estimate the total body-burden for workers who deal with unsealed radioactive isotopes that may give rise to levels of ingestion or inhalation in excess of the maximum permissible concentrations. Such tests should also be performed where radioisotopes may enter the body through the skin or through skin punctures and open wounds. These tests may require the monitoring of breath and of excreta, and the direct determination of the body-burden by means of a total body monitor, according to circumstances. The radiation doses delivered to the appropriate organs or tissues should be calculated and noted on the personal record, and the permitted doses of external radiation should be adjusted to allow for the " internal " doses.

*Addition of Doses from External and Internal Exposures*

(86a) The Commission has considered the difficulty of making proper allowance for the addition of doses from penetrating external exposure and from internal exposure *in the case of long-lived bone-seeking isotopes*, in view of the use for them of the " $n$ " factor. For these radionuclides the Commission recommends that

(i) if the estimated body-burden is less than one-half of the maximum permissible, no consequential restriction of external radiation exposure need be applied ;

(ii) if the estimated body-burden is greater than one-half but less than the maximum permissible, the total body exposure to penetrating external radiation shall be limited to not more than 1.5 rems in any year ; and if $A$ years is the age of the individual when the body-burden was first found to exceed one-half the maximum permissible, then the total accumulated dose from such future exposure beyond age $A$ shall not exceed $1.5(N - A)$ rems where $N$ is the current age of the individual ; if, however, at age $B$ the body-burden is found to have dropped to less than one-half the maximum permissible value, the accumulated dose from future external total body exposure beyond age $B$ shall not exceed $5(N - B)$ rems.

(iii) if the estimated body-burden equals or exceeds the maximum permissible, no occupational exposure to penetrating external radiation is permissible.

### RADIATION WARNING DEVICES

(87) An appropriate form of warning shall be provided to indicate the existence of a radiation hazard, even if the hazard is of a temporary nature only. In the latter case the warning device should be removed when the hazard no longer exists.